Mathematics Standard Level
for the IB Diploma

Solutions Manual

Paul Fannon, Vesna Kadelburg,
Ben Woolley and Stephen Ward

CAMBRIDGE
UNIVERSITY PRESS

University Printing House, Cambridge CB2 8BS, United Kingdom

One Liberty Plaza, 20th Floor, New York, NY 10006, USA

477 Williamstown Road, Port Melbourne, VIC 3207, Australia

4843/24, 2nd Floor, Ansari Road, Daryaganj, Delhi – 110002, India

79 Anson Road, #06-04/06, Singapore 079906

Cambridge University Press is part of the University of Cambridge.

It furthers the University's mission by disseminating knowledge in the pursuit of education, learning and research at the highest international levels of excellence.

www.cambridge.org
Information on this title: education.cambridge.org

© Cambridge University Press 2016

This publication is in copyright. Subject to statutory exception and to the provisions of relevant collective licensing agreements, no reproduction of any part may take place without the written permission of Cambridge University Press.

First published 2016
20 19 18 17 16 15 14 13 12 11 10 9 8 7 6 5 4 3

Printed in Great Britain by CPI Group (UK) Ltd, Croydon CR0 4YY

A catalogue record for this publication is available from the British Library

ISBN 978-1-107-57924-8 Paperback

Cover image: David Robertson/Alamy

Cambridge University Press has no responsibility for the persistence or accuracy of URLs for external or third-party internet websites referred to in this publication, and does not guarantee that any content on such websites is, or will remain, accurate or appropriate. Information regarding prices, travel timetables, and other factual information given in this work is correct at the time of first printing but Cambridge University Press does not guarantee the accuracy of such information thereafter.

NOTICE TO TEACHERS IN THE UK
It is illegal to reproduce any part of this work in material form (including photocopying and electronic storage) except under the following circumstances:

(i) where you are abiding by a licence granted to your school or institution by the Copyright Licensing Agency;
(ii) where no such licence exists, or where you wish to exceed the terms of a licence, and you have gained the written permission of Cambridge University Press;
(iii) where you are allowed to reproduce without permission under the provisions of Chapter 3 of the Copyright, Designs and Patents Act 1988, which covers, for example, the reproduction of short passages within certain types of educational anthology and reproduction for the purposes of setting examination questions.

Contents

	Introduction	iv
Chapter 1	Quadratic functions	1
Chapter 2	Exponents and logarithms	10
Chapter 3	Algebraic structures	22
Chapter 4	The theory of functions	27
Chapter 5	Transformations of graphs	39
Chapter 6	Sequences and series	45
Chapter 7	Binomial expansion	60
Chapter 8	Circular measure and trigonometric functions	69
Chapter 9	Trigonometric equations and identities	75
Chapter 10	Geometry of triangles and circles	87
Chapter 11	Vectors	103
Chapter 12	Basic differentiation and its applications	120
Chapter 13	Basic integration and its applications	136
Chapter 14	Further differentiation	144
Chapter 15	Further integration	159
Chapter 16	Summarising data	169
Chapter 17	Probability	180
Chapter 18	Probability distributions	203
Chapter 19	Questions crossing chapters	215

Introduction

This book contains worked solutions for all the exam-style questions that, in the Standard Level coursebook, are colour-coded green, blue, red or gold. As a reminder:

Green questions should be accessible to students on the way to achieving a grade 3 or 4.

Blue questions are aimed at students hoping to attain a grade 5 or 6.

Red questions are the most difficult, and success with a good number of these would suggest a student is on course for a grade 7.

Gold questions are of a type not typically seen in the exam but are designed to provoke thinking and discussion in order to develop a better understanding of the topic.

Of course, these are just guidelines. If you are aiming for a grade 6, do not be surprised if occasionally you find a green question you cannot do; people are rarely equally good at all areas of the syllabus. Similarly, even if you are able to do all the red questions, that does not guarantee you will get a grade 7 – after all, in the examination you will have to deal with time pressure and exam stress! It is also worth remembering that these questions are graded according to our experience of the final examination. When you first start the course, you may well find the questions harder than you would do by the end of the course, so try not to get discouraged!

The solutions are generally written in the format of a 'model answer' that students should aim to produce in the exam, but sometimes extra lines of working are included (which wouldn't be absolutely necessary in the exam) in order to make the solution easier to follow and to aid understanding.

In many cases the approach shown is not the only option, and neither do we claim it to always be categorically the best or the approach that we would advise every student to take; this is clearly subjective, and different students will have different strengths and therefore preferences. Alternative methods are sometimes given, either in the form of a full worked solution or by way of a comment after the given worked solution.

Where the question has a calculator symbol next to it (indicating that it is for the calculator paper), the approach taken is intentionally designed to utilise the calculator and thereby to minimise work. Students should make sure they are familiar with all the calculator techniques used and, if not, the calculator skills sheets on the CD-ROM accompanying the coursebook should be consulted.

When there is no symbol (indicating that the question could appear on either the calculator or the non-calculator paper), the solution given usually assumes that a calculator is not available. Students should make sure they can cope in this situation but also that they can adapt and use features of the calculator to speed up the process. Perhaps the most common example of this is to use the calculator's equation solver rather than solving simultaneous or quadratic equations manually.

We strongly advise that these solutions be consulted only after having spent a good deal of time and effort working on a question. The process of thinking about the problems encountered, even if a full solution cannot ultimately be found, is really important in developing the skills and knowledge needed to succeed in the future.

We hope you find these solutions useful and that, when used wisely in conjunction with the coursebook, they lead to success in the IB exam.

Stephen Ward, Paul Fannon, Vesna Kadelburg, Ben Woolley

1 Quadratic functions

Exercise 1B

4 a $y = x^2 - 6x + 11$
$= (x-3)^2 - 9 + 11$
$= (x-3)^2 + 2$

b Minimum value of y is 2.

5 a Minimum at $(3, 6) \Rightarrow y = a(x-3)^2 + 6$
So $a = -3$, $b = 6$

b Curve passes through $(1, 14)$, so substituting $x = 1$ and $y = 14$ into $y = a(x-3)^2 + 6$:

$14 = a(1-3)^2 + 6$
$8 = 4a$
$a = 2$

6 a $2x^2 + 4x - 1 = 2\left[x^2 + 2x\right] - 1$
$= 2\left[(x+1)^2 - 1\right] - 1$
$= 2(x+1)^2 - 2 - 1$
$= 2(x+1)^2 - 3$

b Line of symmetry is $x = -1$

c $2x^2 + 4x - 1 = 0$
$2(x+1)^2 - 3 = 0$
$2(x+1)^2 = 3$
$(x+1)^2 = \dfrac{3}{2}$
$x + 1 = \pm\sqrt{\dfrac{3}{2}}$
$x = -1 \pm \sqrt{\dfrac{3}{2}}$

Exercise 1C

4 a $2x^2 + 5x - 12 = (2x - 3)(x + 4)$

b The graph crosses the x-axis where $y = 0$, i.e. where $2x^2 + 5x - 12 = 0$:
$2x^2 + 5x - 12 = 0$
$(2x - 3)(x + 4) = 0$
$2x - 3 = 0$ or $x = 4 = 0$
$x = \dfrac{3}{2}$ or $x = -4$

So the intersections with the x-axis are at $\left(\dfrac{3}{2}, 0\right)$ and $(-4, 0)$.

5 Roots at -5 and 2
$\Rightarrow y = a(x-2)(x+5)$
$= ax^2 + 3ax - 10a$
So $c = -10a$ and $b = 3a$.

y-intercept at $3 \Rightarrow c = 3$
$\therefore 3 = -10a$
$a = -\dfrac{3}{10}$
and $b = 3\left(-\dfrac{3}{10}\right) = -\dfrac{9}{10}$

Exercise 1D

5 $3x^2 = 4x + 1 \Leftrightarrow 3x^2 - 4x - 1 = 0$
Using the quadratic formula with $a = 3$, $b = -4$ and $c = -1$:

$$x = \frac{-(-4) \pm \sqrt{(-4)^2 - 4 \times 3 \times (-1)}}{2 \times 3}$$

$$= \frac{4 \pm \sqrt{28}}{6}$$

$$= \frac{2 \pm \sqrt{7}}{3}$$

6 The discriminant with $a = 4$, $b = 1$ and $c = \dfrac{1}{16}$ is

$$\Delta = b^2 - 4ac$$

$$= 1^2 - 4 \times 4 \times \frac{1}{16}$$

$$= 0$$

So there is only one root, i.e. the vertex lies on the x-axis.

7 Equal roots when discriminant is zero:

$$\Delta = b^2 - 4ac = 0$$

$$(-4)^2 - 4 \times m \times 2m = 0$$

$$16 - 8m^2 = 0$$

$$m^2 = 2$$

$$m = \pm\sqrt{2}$$

8 Tangent to the x-axis implies equal roots, so discriminant is zero:

$$\Delta = b^2 - 4ac = 0$$

$$(2k+1)^2 - 4 \times (-3) \times (-4k) = 0$$

$$4k^2 + 4k + 1 - 48k = 0$$

$$4k^2 - 44k + 1 = 0$$

$$k = \frac{44 \pm \sqrt{44^2 - 4 \times 4 \times 1}}{2 \times 4}$$

$$= \frac{44 \pm \sqrt{1920}}{8}$$

$$= \frac{11}{2} \pm \sqrt{30}$$

9 No real solutions when discriminant $\Delta < 0$:

$$b^2 - 4ac < 0$$

$$(-6)^2 - 4 \times 1 \times 2k < 0$$

$$36 - 8k < 0$$

$$k > \frac{9}{2}$$

10 For a quadratic to be non-negative (≥ 0) for all x, it must have at most one root, so $\Delta \leq 0$ and $a > 0$.

$$b^2 - 4ac \leq 0$$

$$(-3)^2 - 4 \times 2 \times (2c - 1) \leq 0$$

$$9 - 16c + 8 \leq 0$$

$$c \geq \frac{17}{16}$$

COMMENT

Note that $\Delta \leq 0$ is not sufficient in general for a quadratic to be non-negative. The condition $a > 0$ is also necessary to ensure that the quadratic has a positive shape (opening upward) rather than a negative shape (opening downward), so that the curve remains above the x-axis and never goes below it, as would be the case if $a < 0$. In this question a was given as positive (2), so we did not need to use this condition at all.

11 For a quadratic to be negative for all x, it must have no real roots, so $\Delta < 0$ and $a < 0$.

$$b^2 - 4ac < 0$$

$$3^2 - 4 \times m \times (-4) < 0$$

$$9 + 16m < 0$$

$$m < -\frac{9}{16}$$

COMMENT

The condition $a < 0$ ensures that the function is negative shaped and therefore remains below the x-axis. In this case $a = m$, and it followed from the condition on Δ that $a < 0$, as seen in the answer.

12 The two zeros of $ax^2 + bx + c$ are
$$\frac{-b+\sqrt{b^2-4ac}}{2a} \text{ and } \frac{-b-\sqrt{b^2-4ac}}{2a}.$$
The positive difference between these zeros is
$$\left|\frac{-b+\sqrt{b^2-4ac}}{2a} - \frac{-b-\sqrt{b^2-4ac}}{2a}\right| = \left|\frac{2\sqrt{b^2-4ac}}{2a}\right|$$
$$= \left|\frac{\sqrt{b^2-4ac}}{a}\right|$$

So, in this case,
$$\frac{\sqrt{k^2-12}}{1} = \sqrt{69}$$
$k^2 - 12 = 69$
$k^2 = 81$
$k = \pm 9$

COMMENT

Note that modulus signs were used in the general expression for the positive distance, as a could be negative. Here $a = 1$ and so the modulus was not required in the specific case in this question.

Exercise 1E

3 $y = x^2 - 4$ ...(1)
$y = 8 - x$ ...(2)

Substituting (1) into (2):
$x^2 - 4 = 8 - x$
$x^2 + x - 12 = 0$
$(x-3)(x+4) = 0$
$x = 3$ or $x = -4$

Substituting into (2):
$x = 3$: $y = 8 - 3 = 5$
$x = -4$: $y = 8 - (-4) = 12$

So the points of intersection are (3, 5) and (−4, 12).

4 $y = 2x^2 - 3x + 2$ ...(1)
$3x + 2y = 5$ ...(2)

Substituting (1) into (2):
$3x + 2(2x^2 - 3x + 2) = 5$
$4x^2 - 3x - 1 = 0$
$(4x+1)(x-1) = 0$
$x = -\frac{1}{4}$ or $x = 1$

Substituting into (1):
$x = -\frac{1}{4}$: $y = 2\left(-\frac{1}{4}\right)^2 - 3\left(-\frac{1}{4}\right) + 2 = \frac{23}{8}$
$x = 1$: $y = 2 \times 1^2 - 3 \times 1 + 2 = 1$

So the solutions are $\left(-\frac{1}{4}, \frac{23}{8}\right)$ and (1, 1).

5 a $x^2 - 6x + y^2 - 2y - 8 = 0$ ...(1)
$y = x - 8$ ...(2)

Substituting (2) into (1):
$x^2 - 6x + (x-8)^2 - 2(x-8) - 8 = 0$
$2x^2 - 24x + 72 = 0$
$x^2 - 12x + 36 = 0$
as required.

b $x^2 - 12x + 36 = 0$
$(x-6)^2 = 0$
$x = 6$

1 Quadratic functions

There is only one point of intersection, which means that the line is tangent to the circle.

6 $y = mx + 3$ ...(1)

$y = 3x^2 - x + 5$ ...(2)

Substituting (1) into (2):

$mx + 3 = 3x^2 - x + 5$

$3x^2 - x - mx + 2 = 0$

$3x^2 - (m+1)x + 2 = 0$

Only one intersection means that this quadratic has a single root, so $\Delta = 0$:

$b^2 - 4ac = 0$

$[-(m+1)]^2 - 4 \times 3 \times 2 = 0$

$(m+1)^2 = 24$

$m + 1 = \pm\sqrt{24}$

$m = -1 \pm \sqrt{24} = -1 \pm 2\sqrt{6}$

> **COMMENT**
>
> It is often wise to convert decimals (such as 9.75) into fractions to make subsequent manipulation easier. The quadratic could also have been solved with a GDC, of course.

Exercise 1F

1 Let one number be x and the other be y.

Sum of x and y is 8: $x + y = 8$ ...(1)

Product is 9.75: $xy = 9.75$ ...(2)

From (1): $y = 8 - x$ ...(3)

Substituting (3) into (2):

$x(8 - x) = 9.75$

$x^2 - 8x + \dfrac{39}{4} = 0$

$4x^2 - 32x + 39 = 0$

$(2x - 3)(2x - 13) = 0$

$x = \dfrac{3}{2}$ or $x = \dfrac{13}{2}$

The two numbers are 1.5 and 6.5.

2 The length is x; let the width be y. Perimeter of 12:

$2(x + y) = 12$

$x + y = 6$

$y = 6 - x$

Area $A = xy$

$= x(6 - x)$

$= 6x - x^2$

Completing the square:

$A = -[x^2 - 6x]$

$= -(x - 3)^2 + 9$

So the maximum area is $9\,\text{cm}^2$ (and occurs when $x = y = 3$, i.e. when the rectangle is a square with side 3 cm).

> **COMMENT**
>
> Note that the negative sign makes the quadratic negative shaped, which results in a maximum rather than minimum turning point.

3 a New fencing required is 200 m, so

$2x - 10 + y = 200$

$\Rightarrow y = 210 - 2x$

Area $A = xy$

$= x(210 - 2x)$

$= 210x - 2x^2$

b Completing the square:

$$A = -2\left[x^2 - 105x\right]$$

$$= -2\left[\left(x - \frac{105}{2}\right)^2 - \left(\frac{105}{2}\right)^2\right]$$

$$= 2\left(x - \frac{105}{2}\right)^2 - \frac{105^2}{2}$$

∴ maximum area when

$$x = \frac{105}{2} = 52.5\,\text{m}, \text{ for which}$$

$$y = 210 - 2 \times \frac{105}{2} = 105\,\text{m}$$

4 a Ball is at ground level when $h = 0$.

$$8t - 4.9t^2 = 0$$

$$t(8 - 4.9t) = 0$$

$$t = 0 \quad \text{or} \quad t = \frac{8}{4.9} = 1.63\,\text{(3SF)}$$

So the ball returns to the ground after 1.63 s.

b By symmetry of a quadratic, maximum height (vertex) is halfway between the roots, $t = 0$ and $t = \frac{8}{4.9}$.

$$t_{max} = \frac{0 + \frac{8}{4.9}}{2} = \frac{4}{4.9}$$

$$\therefore h_{max} = 8\left(\frac{4}{4.9}\right) - 4.9\left(\frac{4}{4.9}\right)^2$$

$$= \frac{32 - 16}{4.9} = \frac{16}{4.9} = 3.27\,\text{m}\,\text{(3SF)}$$

> **COMMENT**
>
> This could also have been solved using a GDC or by completing the square.

5 a Perimeter of 60:

$$2x + \frac{1}{2}\pi y = 60$$

$$\Rightarrow 2x = 60 - \frac{\pi y}{2}$$

$$\Rightarrow x = 30 - \frac{\pi y}{4}$$

$$A = xy + \frac{1}{2}\pi\left(\frac{y}{2}\right)^2$$

$$= \left(30 - \frac{\pi y}{4}\right)y + \frac{\pi y^2}{8}$$

$$= 30y - \frac{\pi y^2}{4} + \frac{\pi y^2}{8}$$

$$= 30y - \frac{1}{8}\pi y^2$$

b Finding the roots of the area expression:

$$\left(30y - \frac{1}{8}\pi y^2\right) = 0$$

$$y\left(30 - \frac{1}{8}\pi y\right) = 0$$

$$y = 0 \quad \text{or} \quad y = \frac{240}{\pi}$$

By the symmetry of a quadratic, the maximum area (vertex) is halfway between the roots:

$$y = \frac{0 + \frac{240}{\pi}}{2} = \frac{120}{\pi}$$

When $y = \frac{120}{\pi}$,

$$x = 30 - \frac{\pi y}{4}$$

$$= 30 - \frac{\pi}{4}\left(\frac{120}{\pi}\right)$$

$$= 30 - 30$$

$$= 0$$

> **COMMENT**
> This could also have been solved using a GDC or by completing the square.

c If $A = 200$,

$$30y - \frac{1}{8}\pi y^2 = 200$$

$$\frac{\pi y^2}{8} - 30y + 200 = 0$$

Using the quadratic formula with $a = \frac{\pi}{8}$, $b = -30$ and $c = 200$:

$$y = \frac{30 \pm \sqrt{30^2 - 4 \times \frac{\pi}{8} \times 200}}{\frac{\pi}{4}}$$

$= 7.38$ or 69.0(3SF)

When $y = 7.38$, $x = 30 - \frac{\pi \times 7.38}{4} = 24.2$

When $y = 69.0$, $x = 30 - \frac{\pi \times 69.0}{4} = -24.2$

(therefore reject as $x < 0$)

So $x = 24.2$ m and $y = 7.38$ m.

6 Total profit $= n(200 - 4n) = 4n(50 - n)$

Finding the roots of the total profit function:

$4n(50 - n) = 0$

$n = 0$ or $n = 50$

By the symmetry of a quadratic, the maximum lies halfway between the roots, i.e. at $n = 25$.

> **COMMENT**
> This could also have been solved by completing the square.

Mixed examination practice 1
Short questions

1 a $x^2 + 5x - 14 = (x+7)(x-2)$

b $x^2 + 5x - 14 = 0$
$(x+7)(x-2) = 0$
$x = -7$ or $x = 2$

2 a Positive quadratic, so the vertex is a minimum point.

b Minimum at $(3, 7) \Rightarrow y = (x-3)^2 + 7$
So $a = 3$, $b = 7$

3 Maximum y-value is $48 \Rightarrow c = 48$.
Passes through $(-2, 0)$ and $(6, 0)$ means that its roots are $x = -2$ and $x = 6$. The line of symmetry is midway between the roots, i.e. at $x = 2$, so $b = 2$.

Substituting $x = -2$ and $y = 0$ into $y = a(x-2)^2 + 48$:

$0 = a(-2-2)^2 + 48$
$0 = 16a + 48$
$a = -3$
So $a = -3$, $b = 2$ and $c = 48$.

4 Roots at $x = k$ and $x = k + 4 \Rightarrow$ line of symmetry is $x = k + 2$ (midway between the roots).
So the x-coordinate of the turning point is $k + 2$.

5 a Roots at $-\frac{1}{2}$ and 2, so

$$f(x) = \left(x + \frac{1}{2}\right)(x - 2)$$

i.e. $p = -\frac{1}{2}$, $q = 2$

b Line of symmetry is midway between

the roots: $x = \dfrac{2 + \left(-\dfrac{1}{2}\right)}{2} = \dfrac{3}{4}$

∴ x-coordinate of C is $\dfrac{3}{4}$

6
- Negative quadratic $\Rightarrow a$ is negative
- Negative y-intercept $\Rightarrow c$ is negative
- Single (repeated) root $\Rightarrow b^2 - 4ac = 0$
- Line of symmetry $x = -\dfrac{b}{2a}$ is positive

  $\Rightarrow b$ is positive (as a is negative)

TABLE 1MS.6

Expression	Positive	Negative	Zero
a		✓	
c		✓	
$b^2 - 4ac$			✓
b	✓		

7 a $x^2 - 10x + 35 = (x-5)^2 - 25 + 35$

$= (x-5)^2 + 10$

b From (a), the minimum value of $x^2 - 10x + 35$ is 10.

Hence the maximum value of

$\dfrac{1}{(x^2 - 10x + 35)^3}$ is $\dfrac{1}{10^3} = \dfrac{1}{1000}$

8 Equal roots $\Rightarrow \Delta = 0$

$b^2 - 4ac = 0$

$(k+1)^2 - 4 \times 2k \times 1 = 0$

$k^2 - 6k + 1 = 0$

$k = \dfrac{6 \pm \sqrt{32}}{2} = 3 \pm 2\sqrt{2}$

9 No real roots $\Rightarrow \Delta < 0$

$b^2 - 4ac < 0$

$6^2 - 4 \times 2 \times k < 0$

$36 - 8k < 0$

$k > \dfrac{9}{2}$

10 Only one zero $\Rightarrow \Delta = 0$

$b^2 - 4ac = 0$

$[-(k+1)]^2 - 4 \times 1 \times 3 = 0$

$(k+1)^2 - 12 = 0$

$k + 1 = \pm 2\sqrt{3}$

$k = -1 \pm 2\sqrt{3}$

11 a Roots of $x^2 - kx + (k-1) = 0$ are

$\dfrac{k \pm \sqrt{k^2 - 4(k-1)}}{2} = \dfrac{k \pm \sqrt{k^2 - 4k + 4}}{2}$

$= \dfrac{k \pm \sqrt{(k-2)^2}}{2}$

$= \dfrac{k \pm (k-2)}{2}$

$= k - 1 \text{ or } 1$

∴ $\alpha = k-1, \ \beta = 1$

b $\alpha^2 + \beta^2 = 17$

$(k-1)^2 + 1 = 17$

$k^2 - 2k + 2 = 17$

$k^2 - 2k - 15 = 0$

$(k-5)(k+3) = 0$

$k = 5$ or $k = -3$

1 Quadratic functions 7

Long questions

1 a i Square perimeter $= 4x$

  ii Circle perimeter $= 2\pi y$

b $4x + 2\pi y = 8 \Rightarrow x = 2 - \dfrac{\pi}{2} y$

c $A =$ area of square + area of circle

$$= x^2 + \pi y^2$$

$$= \left(2 - \dfrac{\pi y}{2}\right)^2 + \pi y^2$$

$$= 4 - 2\pi y + \dfrac{\pi^2 y^2}{4} + \pi y^2$$

$$= \dfrac{\pi}{4}(\pi + 4) y^2 - 2\pi y + 4$$

d Completing the square:

$$A = \dfrac{\pi}{4}(\pi+4)\left[y^2 - \dfrac{8}{\pi+4} y + \dfrac{16}{\pi(\pi+4)} \right]$$

$$= \dfrac{\pi}{4}(\pi+4)\left[\left(y - \dfrac{4}{\pi+4}\right)^2 - \left(\dfrac{4}{\pi+4}\right)^2 + \dfrac{16}{\pi(\pi+4)} \right]$$

So the minimum area occurs when $y = \dfrac{4}{\pi+4}$

Percentage of wire in circle

$$= \dfrac{\text{length of wire in circle}}{\text{total length of wire}} \times 100\%$$

$$= \dfrac{2\pi y}{8} \times 100\%$$

$$= \dfrac{2\pi \left(\dfrac{4}{\pi+4}\right)}{8} \times 100\%$$

$$= 44.0\% \, (3\text{SF})$$

> **COMMENT**
>
> Note that it isn't necessary to simplify the constant in the expression for A after completing the square, as the question asks only for the value of y where the area is minimised and not for the actual value of that minimum.

2 a Car A has position $(20t - 50, 0)$ and Car B has position $(0, 15t - 30)$.

$$d^2 = (x_2 - x_1)^2 + (y_2 - y_1)^2$$
$$= [0-(20t-50)]^2 + [(15t-30)-0]^2$$
$$= (20t-50)^2 + (15t-30)^2$$
$$= 400t^2 - 2000t + 2500 + 225t^2 - 900t + 900$$
$$= 625t^2 - 2900t + 3400$$

b Completing the square:

$$d^2 = 625\left[t^2 - \frac{116}{25}t\right] + 3400$$
$$= 625\left[\left(t - \frac{58}{25}\right)^2 - \left(\frac{58}{25}\right)^2\right] + 3400$$
$$= 625\left(t - \frac{58}{25}\right)^2 - 58^2 + 3400$$
$$= 625\left(t - \frac{58}{25}\right)^2 + 36$$

So $d^2 \geq 36$ and, since $d > 0$, it follows that the minimum value of d is 6 km.

3 a Vertex on the x-axis
$\Rightarrow$ has only one root, so $\Delta = 0$.

$$b^2 - 4ac = 0$$
$$36 - 4k = 0$$
$$k = 9$$

b Equation of first graph is

$$y = x^2 - 6x + 9 = (x-3)^2$$

So vertex is at $(3, 0)$.
Second graph has vertex at $(-2, 5)$, so its equation is $y = a(x+2)^2 + 5$

It passes through $(3, 0)$; substituting into the equation gives

$$0 = a(3+2)^2 + 5$$
$$25a = -5$$
$$a = -\frac{1}{5}$$

$$\therefore y = -\frac{1}{5}(x+2)^2 + 5$$
$$= -\frac{1}{5}(x^2 + 4x + 4) + 5$$
$$= -\frac{1}{5}x^2 - \frac{4}{5}x + \frac{21}{5}$$

c For intersection of $y = x^2 - 6x + 9$ and $y = -\frac{1}{5}x^2 - \frac{4}{5}x + \frac{21}{5}$:

$$x^2 - 6x + 9 = -\frac{1}{5}x^2 - \frac{4}{5}x + \frac{21}{5}$$
$$5x^2 - 30x + 45 = -x^2 - 4x + 21$$
$$6x^2 - 26x + 24 = 0$$
$$3x^2 - 13x + 12 = 0$$
$$(3x-4)(x-3) = 0$$
$$x = \frac{4}{3} \text{ or } x = 3$$

$x = 3$ is the point of intersection at the vertex $(3, 0)$ of the first graph.

To find the y-coordinate of the other point, substitute $x = \frac{4}{3}$ into $y = (x-3)^2$:

$$y = \left(\frac{4}{3} - 3\right)^2 = \frac{25}{9}$$

So the other point of intersection is $\left(\frac{4}{3}, \frac{25}{9}\right)$.

1 Quadratic functions 9

2 Exponents and logarithms

Exercise 2A

10 n inputs are sorted in $k \times n^{1.5}$ microseconds.

1 million = 10^6 inputs are sorted in 0.5 seconds = 0.5×10^6 microseconds.

$$\therefore k \times (10^6)^{1.5} = 0.5 \times 10^6$$
$$k \times 10^9 = 0.5 \times 10^6$$
$$k = 0.5 \times 10^{-3}$$
$$= 5 \times 10^{-4}$$

11 $V = xy^2$, and when $x = 2y$, $V = 128$

$$\therefore (2y) \times y^2 = 128$$
$$y^3 = 64$$
$$y = 4$$

Hence $x = 2 \times 4 = 8$ cm.

12 a Substituting $A = 81$ and $V = 243$:

$$V = kA^{1.5}$$
$$243 = k \times (81)^{\frac{3}{2}}$$
$$243 = k \times (81^{\frac{1}{2}})^3$$
$$243 = k \times 9^3$$
$$243 = 729k$$
$$k = \frac{1}{3}$$

b If $V = \dfrac{64}{3}$, then

$$\frac{64}{3} = \frac{1}{3} A^{1.5}$$
$$A^{\frac{3}{2}} = 64$$
$$A = 64^{\frac{2}{3}}$$
$$= (64^{\frac{1}{3}})^2$$
$$= 4^2 = 16 \text{ cm}^2$$

13 $2 \times 5^{x-1} = 250$
$$5^{x-1} = 125$$
$$5^{x-1} = 5^3$$
$$x - 1 = 3$$
$$x = 4$$

14 $5 + 3^{x+2} = 14$
$$3^{x+2} = 9$$
$$3^{x+2} = 3^2$$
$$x + 2 = 2$$
$$x = 0$$

15 $100^{x+5} = 10^{3x-1}$
$$(10^2)^{x+5} = 10^{3x-1}$$
$$10^{2x+10} = 10^{3x-1}$$
$$2x + 10 = 3x - 1$$
$$x = 11$$

16
$$16 + 2^x = 2^{x+1}$$
$$2^4 + 2^x = 2^{x+1}$$
$$2^4 + 2^x = 2 \times 2^x$$
$$2^4 = 2 \times 2^x - 2^x$$
$$2^4 = 2^x$$
$$x = 4$$

17
$$6^{x+1} = 162 \times 2^x$$
$$6 \times 6^x = 162 \times 2^x$$
$$\frac{6^x}{2^x} = \frac{162}{6}$$
$$\left(\frac{6}{2}\right)^x = 27$$
$$3^x = 27$$
$$x = 3$$

18
$$4^{1.5x} = 2 \times 16^{x-1}$$
$$(4^{\frac{3}{2}})^x = 2 \times \frac{1}{16} \times 16^x$$
$$8^x = \frac{1}{8} \times 16^x$$
$$8^{x+1} = 16^x$$
$$(2^3)^{x+1} = (2^4)^x$$
$$2^{3x+3} = 2^{4x}$$
$$3x + 3 = 4x$$
$$x = 3$$

Exercise 2B

2 At 09:00 on Tuesday, $t = 0$ and $y = 10$.
Substituting:
$$y = k \times 1.1^t$$
$$10 = k \times 1.1^0$$
$$k = 10$$
At 09:00 on Friday, $t = 3$
$$\therefore y = 10 \times 1.1^3 = 13.31 \text{ m}^2$$

3 $h = 2 - 0.2 \times 1.6^{0.2m}$

a

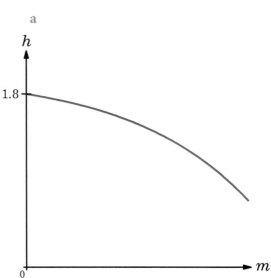

Figure 2B.3 Graph of $h = 2 - 0.2 \times 1.6^{0.2m}$

b When there is no fruit, $m = 0$, so
$h = 2 - 0.2 \times 1.6^0 = 1.8$ m

c When $m = 7.5$,
$h = 2 - 0.2 \times 1.6^{0.2 \times 7.5} = 1.60$ m (3SF)

d The model was derived from data which, from (c), gave a height of 1.6 m above the ground at the harvest-time fruit load of 7.5 kg. Extrapolating so far beyond the model to reach $h = 0$ is unreliable and likely to be unrealistic; for example, the branch might simply break before being bent far enough to touch the ground.

4 a $y = 1 + 16^{1-x^2} = 1 + \dfrac{1}{16^{x^2-1}}$

As $x \to \infty$, $\dfrac{1}{16^{x^2-1}} \to 0$ and so $y \to 1$.

The maximum value for y must occur at the minimum value of $x^2 - 1$, which is when $x = 0$.

When $x = 0$, $y = 1 + \dfrac{1}{16^{0-1}} = 17$, so the maximum point is (0, 17).

2 Exponents and logarithms 11

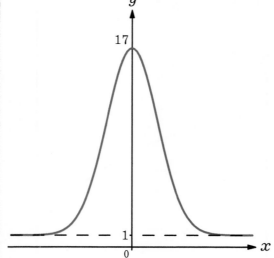

Figure 2B.4 Graph of $y = 1 + 16^{1-x^2}$

> **COMMENT**
> Some justification of how this graph could be constructed is given here, but you could just draw it on a GDC.

  b When $y = 3$:

$$3 = 1 + 16^{1-x^2}$$
$$2 = 16^{1-x^2}$$
$$16^{\frac{1}{4}} = 16^{1-x^2}$$
$$\frac{1}{4} = 1 - x^2$$
$$x^2 = \frac{3}{4}$$
$$x = \pm\frac{\sqrt{3}}{2} = \pm 0.866 \,(3\text{SF})$$

5 Temperature T of the soup decreases exponentially with time towards 20 (room temperature):

$$T = 20 + A \times m^{kt}$$

When the soup is served, $t = 0$ and $T = 55$

$$\therefore 55 = 20 + A \times m^0$$
$$A = 35$$

Every 5 minutes the term $A \times m^{kt}$ must decrease by 30%, so $m = 0.7$ and $k = \frac{1}{5} = 0.2$.

$$\therefore T = 20 + 35 \times 0.7^{0.2t}$$

When $t = 7$,

$$T = 20 + 35 \times 0.7^{1.4} = 41.2°C \,(3\text{SF})$$

6 $V = 40(1 - 3^{-0.1t})$

  a When $t = 0$,

$$V = 40(1-1) = 0 \text{ m s}^{-1}$$

  b As $t \to \infty$, $3^{-0.1t} \to 0$ and so

$$V \to 40(1-0) = 40 \text{ m s}^{-1}$$

7 $T = A + B \times 2^{-\frac{x}{y}}$

  a Background temperature is 25°C, so as $x \to \infty$, $T \to 25$.

Since $2^{-\frac{x}{k}} \to 0$, $T \to A$.

Hence $A = 25$

Temperature on surface of light bulb is 125°C, so when $x = 0$, $T = 125$. Substituting:

$$125 = 25 + B \times 2^0$$
$$\Rightarrow B = 100$$

Air temperature 3 mm from surface of light bulb is 75°C, so when $x = 3$, $T = 75$. Substituting:

$$75 = 25 + 100 \times 2^{-\frac{3}{k}}$$
$$\frac{1}{2} = 2^{-\frac{3}{k}}$$
$$2^{\frac{3}{k}} = 2$$
$$\frac{3}{k} = 1$$
$$k = 3$$

b At 2 cm from the surface of the bulb, $x = 20$ (mm), so

$$T = 25 + 100 \times 2^{-\frac{20}{3}}$$
$$= 26.0°C$$

c

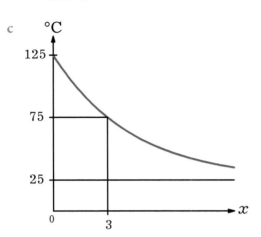

Figure 2B.7 Graph of $T = 25 + 100 \times 2^{-\frac{x}{3}}$

Exercise 2C

2 This is a very close approximation to e (with less than a 7.5×10^{-7}% error).

> **COMMENT**
>
> It might at first seem that this is far too unlikely to be a mere coincidence and that there must be some underlying relationship, but this is in fact not the case. The so-called 'Strong Law of Small Numbers' gives some insight into the surprisingly regular occurrence of this type of coincidence.

3 $\left(e^2 + \dfrac{2}{e^2}\right)^2 = e^4 + \left(e^2 \times \dfrac{2}{e^2}\right) + \left(\dfrac{2}{e^2} \times e^2\right) + \dfrac{4}{e^4}$

$\qquad = e^4 + 2 + 2 + \dfrac{4}{e^4}$

$\qquad = e^4 + 4 + \dfrac{4}{e^4}$

Exercise 2D

6 $\log_{10}(9x+1) = 3$

$\qquad 9x + 1 = 10^3$

$\qquad 9x + 1 = 1000$

$\qquad x = 111$

7 $\log_8 \sqrt{1-x} = \dfrac{1}{3}$

$\qquad \sqrt{1-x} = 8^{\frac{1}{3}}$

$\qquad \sqrt{1-x} = 2$

$\qquad 1 - x = 4$

$\qquad x = -3$

8 $\ln(3x-1) = 2$

$\qquad 3x - 1 = e^2$

$\qquad x = \dfrac{e^2 + 1}{3}$

9 $(\log_3 x)^2 = 4$

$\qquad \log_3 x = \pm 2$

$\qquad x = 3^{\pm 2} = 9 \text{ or } \dfrac{1}{9}$

10 $3(1 + \log x) = 6 + \log x$

$\qquad 3 + 3\log x = 6 + \log x$

$\qquad 2\log x = 3$

$\qquad \log x = \dfrac{3}{2}$

$\qquad x = 10^{\frac{3}{2}}$

$\qquad = 10\sqrt{10} = 31.6 \, (3\text{SF})$

11 $\log_x 4 = 9$

$\qquad 4 = x^9$

$\qquad x = 4^{\frac{1}{9}} = 1.17 \, (3\text{SF})$

2 Exponents and logarithms

12 $\log_3 x + \log_5 y = 6$ ...(1)
$\log_3 x - \log_5 y = 2$ ...(2)

(1) + (2):

$2\log_3 x = 8$
$\log_3 x = 4$
$x = 3^4 = 81$

Substituting into (1):

$4 + \log_5 y = 6$
$\log_5 y = 2$
$y = 5^2 = 25$

i.e. $x = 81$, $y = 25$

13 Let r be the Richter-scale value and s the strength of an earthquake.
Since an increase of one unit in r corresponds to an increase by a factor of 10 in s,

$s = C \times 10^r$ for some constant C.

Let t be the strength of an earthquake of Richter level 5.2:

$t = C \times 10^{5.2}$ ...(1)

For an earthquake twice as strong:

$2t = C \times 10^r$ ...(2)

(2) ÷ (1): $\dfrac{2t}{t} = \dfrac{C \times 10^r}{C \times 10^{5.2}}$

$2 = 10^{r-5.2}$
$r - 5.2 = \log 2$
$r = \log 2 + 5.2 = 5.50$

An earthquake twice as strong as a level-5.2 quake would measure 5.5 on the Richter scale.

> **COMMENT**
> The constant C is needed here as the precise relationship between r and s is not given, but it is not necessary to find the value of C to answer this particular question.

Exercise 2E

4 $2\ln x + \ln 9 = 3$

$\ln x = \dfrac{3 - \ln 9}{2}$

$= \dfrac{3 - 2\ln 3}{2}$

$= \dfrac{3}{2} - \ln 3$

$\therefore x = e^{\frac{3}{2} - \ln 3}$

$= e^{\frac{3}{2}} \times e^{-\ln 3}$

$= \dfrac{1}{3} e^{\frac{3}{2}}$

5 a $\ln 50 = \ln 2 + \ln 25$
$= \ln 2 + \ln 5^2$
$= \ln 2 + 2\ln 5$
$= a + 2b$

b $\ln 0.16 = \ln\left(\dfrac{4}{25}\right)$

$= \ln\left(\dfrac{2^2}{5^2}\right)$

$= \ln 2^2 - \ln 5^2$
$= 2\ln 2 - 2\ln 5$
$= 2a - 2b$

6 $\log_2 x = \log_x 2$

$\log_2 x = \dfrac{\log_2 2}{\log_2 x}$

$(\log_2 x)^2 = 1$
$\log_2 x = \pm 1$

$x = 2^{\pm 1} = 2$ or $\dfrac{1}{2}$

7 a $\log a^3 - 2\log ab^2$
$= 3\log a - 2(\log a + \log b^2)$
$= 3\log a - 2(\log a + 2\log b)$
$= 3\log a - 2\log a - 4\log b$
$= \log a - 4\log b$
$= x - 4y$

14 Topic 2E Laws of logarithms

b $\log(4b) + 2\log(5ac)$
$= \log 4 + \log b + 2(\log 5 + \log a + \log c)$
$= \log 4 + \log b + 2\log 5 + 2\log a + 2\log c$
$= \log 4 + \log 5^2 + 2x + y + 2z$
$= \log 100 + 2x + y + 2z$
$= 2 + 2x + y + 2z$

8 $\log\frac{1}{2} + \log\frac{2}{3} + \log\frac{3}{4} + \cdots + \log\frac{8}{9} + \log\frac{9}{10}$

$= \log\left(\frac{1}{2} \times \frac{2}{3} \times \frac{3}{4} \times \cdots \times \frac{8}{9} \times \frac{9}{10}\right)$

$= \log\left(\frac{1}{10}\right)$

$= -1$

9 a $\log_a a^2 b = \log_a (a^2) + \log_a b$

$= 2 + \frac{\log b}{\log a}$

$= 2 + \frac{y}{x}$

b $\log_{ab} ac^2 = \frac{\log ac^2}{\log ab}$

$= \frac{\log a + 2\log c}{\log a + \log b}$

$= \frac{x + 2z}{x + y}$

> **COMMENT**
> Remember that if there is no base indicated on a log, then it is assumed that base 10 is being used. In both parts of this question, it was necessary to use the change-of-base rule to introduce a base of 10, i.e. log.

10 a $\log_b\left(\frac{a}{bc}\right) = \log_b a - \log_b bc$

$= \log_b a - (\log_b b + \log_b c)$

$= \log_b a - 1 - \log_b c$

$= \frac{\log a}{\log b} - 1 - \frac{\log c}{\log b}$

$= \frac{x}{y} - 1 - \frac{z}{y}$

$= \frac{x-z}{y} - 1$

b $\log_{a^b}(b^a) = \frac{\log b^a}{\log a^b}$

$= \frac{a \log b}{b \log a}$

$= \frac{10^x y}{10^y x}$

$= \frac{y}{x} \times 10^{x-y}$

Exercise 2G

3 $N = 100e^{1.03t}$

a When $t = 0$,
$N = 100e^0 = 100$

b When $t = 6$,
$N = 100e^{1.03 \times 6} = 48\,300\,(3\text{SF})$

c $N = 1000$ when
$1000 = 100e^{1.03t}$
$e^{1.03t} = 10$
$1.03t = \ln 10$
$t = \frac{1}{1.03}\ln 10$
$= 2.24 \text{ hours}\,(3\text{SF})$

The population will reach 1000 cells after approximately 2 hours and 14 minutes.

> **COMMENT**
>
> 0.24 hours is 0.24 × 60 = 14.4 minutes.

4 a At 9 a.m. 18 people know the rumour, so when $t = 0$, $N = 18$:

$$18 = Ae^0$$
$$A = 18$$

b At 10 a.m. 42 people know the rumour, so when $t = 60$, $N = 42$:

$$42 = 18e^{60k}$$
$$e^{60k} = \frac{42}{18}$$
$$60k = \ln\frac{7}{3}$$
$$k = \frac{1}{60}\ln\frac{7}{3}$$
$$= 0.0141216 = 0.0141 \text{ (3SF)}$$

c At 10:30 a.m., $t = 90$,

$$\therefore N = 18e^{90 \times 0.0141} = 64.2$$

So 64 people know the rumour at 10:30 a.m.

d If 1200 people know the rumour, then

$$1200 = 18e^{0.0141t}$$
$$e^{0.0141t} = \frac{1200}{18}$$
$$0.0141t = \ln\frac{200}{3}$$
$$t = \frac{1}{0.0141}\ln\frac{200}{3} = 297.4$$

So after 298 minutes, i.e. at 13:58, the whole school population will know the rumour.

5 a At $t = 0$,

$$M = ke^0 = k$$

b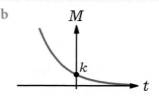

Figure 2G.5 Graph of $M = ke^{-0.01t}$

c 25% of the initial mass means $M = \frac{k}{4}$.

$$\frac{k}{4} = ke^{-0.01t}$$
$$e^{-0.01t} = \frac{1}{4}$$
$$e^{0.01t} = 4$$
$$0.01t = \ln 4$$
$$t = 100\ln 4$$
$$= 138.6 \text{ seconds}$$

The plutonium will be at 25% of its original mass after 2.31 minutes.

> **COMMENT**
>
> 138.6 seconds is 138.6 ÷ 60 = 2.31 minutes.

6 $15^{2x} = 3 \times 5^{x+1}$

$$15^{2x} = 3 \times 5 \times 5^x$$
$$15^{2x} = 15 \times 5^x$$
$$\frac{15^{2x}}{15} = 5^x$$
$$15^{2x-1} = 5^x$$
$$\log 15^{2x-1} = \log 5^x$$
$$(2x-1)\log 15 = x\log 5$$
$$2x\log 15 - \log 15 = x\log 5$$
$$2x\log 15 - x\log 5 = \log 15$$
$$x(2\log 15 - \log 5) = \log 15$$
$$x\log\left(\frac{15^2}{5}\right) = \log 15$$
$$x\log 45 = \log 15$$
$$x = \frac{\log 15}{\log 45}$$

7 $\dfrac{1}{7^x} = 3 \times 49^{5-x}$

$7^{-x} = 3 \times (7^2)^{5-x}$

$7^{-x} = 3 \times 7^{10-2x}$

$\dfrac{7^{-x}}{7^{10-2x}} = 3$

$7^{x-10} = 3$

$x - 10 = \log_7 3$

$x = 10 + \log_7 3$

8 $5 \times 4^{x-1} = \dfrac{1}{3^{2x}}$

$5 \times 4^{-1} \times 4^x = \dfrac{1}{(3^2)^x}$

$\dfrac{5}{4} = \dfrac{\frac{1}{9^x}}{4^x}$

$\dfrac{5}{4} = \dfrac{1}{36^x}$

$36^x = \dfrac{4}{5}$

$\ln 36^x = \ln \dfrac{4}{5}$

$x \ln 36 = \ln \dfrac{4}{5}$

$x = \dfrac{\ln \frac{4}{5}}{\ln 36}$

> **COMMENT**
> The answer in the back of the coursebook,
> $x = \dfrac{\ln\left(\frac{5}{4}\right)}{\ln\left(\frac{1}{36}\right)}$, is an equivalent, though not fully simplified, form.

9 a $y = 3^x$ is always increasing and $y = 3 - x$ is always decreasing, so their two graphs can intersect at most once.

Since $3^0 < 3 - 0$ and $3^1 > 3 - 1$, there must be an intersection in $[0, 1]$.

b From GDC, the intersection occurs at $x = 0.742$ (3SF)

Mixed examination practice 2
Short questions

1 $\log_5\left(\sqrt{x^2 + 49}\right) = 2$

$\sqrt{x^2 + 49} = 25$

$x^2 + 49 = 625$

$x^2 = 576$

$x = \pm 24$

2 a $\log \dfrac{x^2 \sqrt{y}}{z} = \log x^2 + \log \sqrt{y} - \log z$

$= \log x^2 + \log y^{\frac{1}{2}} - \log z$

$= 2\log x + \dfrac{1}{2}\log y - \log z$

$= 2a + \dfrac{b}{2} - c$

b $\log \sqrt{0.1x} = \log(0.1x)^{\frac{1}{2}}$

$= \dfrac{1}{2}\log 0.1x$

$= \dfrac{1}{2}(\log 0.1 + \log x)$

$= \dfrac{1}{2}(-1 + \log x)$

$= \dfrac{a-1}{2}$

c $\log_{100} \dfrac{y}{z} = \dfrac{\log \frac{y}{z}}{\log 100}$

$= \dfrac{\log y - \log z}{\log 100}$

$= \dfrac{b-c}{2}$

3 When $B = 25$,

$$25 = 4 + 12e^{\frac{t}{3}}$$

$$e^{\frac{t}{3}} = \frac{21}{12}$$

$$\frac{t}{3} = \ln\left(\frac{7}{4}\right)$$

$$t = 3\ln\left(\frac{7}{4}\right) = 1.68 \text{ (3SF)}$$

4 $\quad 4 \times 3^{2x} = 5^x$

$$4 \times (3^2)^x = 5^x$$

$$\frac{5^x}{9^x} = 4$$

$$\left(\frac{5}{9}\right)^x = 4$$

$$\ln\left(\frac{5}{9}\right)^x = \ln 4$$

$$x\ln\left(\frac{5}{9}\right) = \ln 4$$

$$x = \frac{\ln 4}{\ln\left(\frac{5}{9}\right)}$$

$$= \frac{\ln 4}{\ln 5 - \ln 9}$$

$$= \frac{2\ln 2}{\ln 5 - 2\ln 3}$$

> **COMMENT**
> The answer in the back of the coursebook, $\frac{\ln 4}{\ln 5 - 2\ln 3}$, is an equivalent, though not fully simplified, form.

5 $\begin{cases} \ln x + \ln y^2 = 8 \\ \ln x^2 + \ln y = 6 \end{cases}$

$\begin{cases} \ln x + 2\ln y = 8 & ...(1) \\ 2\ln x + \ln y = 6 & ...(2) \end{cases}$

$2 \times (2) - (1)$:

$3\ln x = 4$

$$\ln x = \frac{4}{3}$$

$$x = e^{\frac{4}{3}} = 3.79 \text{ (3SF)}$$

Substituting in (2):

$$2 \times \frac{4}{3} + \ln y = 6$$

$$\ln y = \frac{10}{3}$$

$$y = e^{\frac{10}{3}} = 28.0 \text{ (3SF)}$$

6 $\quad y = \ln x - \ln(x+2) + \ln(4-x^2)$

$$= \ln\left(\frac{x(4-x^2)}{x+2}\right)$$

$$= \ln\left(\frac{x(2-x)(2+x)}{x+2}\right)$$

$$= \ln(x(2-x))$$

$$= \ln(2x - x^2)$$

$\therefore e^y = 2x - x^2$

$x^2 - 2x + e^y = 0$

$$x = \frac{2 \pm \sqrt{(-2)^2 - 4 \times 1 \times e^y}}{2}$$

$$= \frac{2 \pm \sqrt{4 - 4e^y}}{2}$$

$$= \frac{2 \pm 2\sqrt{1 - e^y}}{2}$$

$$= 1 \pm \sqrt{1 - e^y}$$

7

$2 \times 3^{x-2} = 36^{x-1}$

$2 \times 3^x \times 3^{-2} = 36^x \times 36^{-1}$

$\dfrac{2}{9} \times 3^x = 36^x \times \dfrac{1}{36}$

$\dfrac{2}{9} \times 36 = \dfrac{36^x}{3^x}$

$8 = \left(\dfrac{36}{3}\right)^x$

$8 = 12^x$

$\ln 8 = \ln 12^x$

$\ln 8 = x \ln 12$

$x = \dfrac{\ln 8}{\ln 12}$

8 Changing $\log_a$ and $\log_b$ into ln:

$\log_a b^2 = c$

$\Rightarrow 2 \log_a b = c$

$\Rightarrow 2 \dfrac{\ln b}{\ln a} = c \quad \ldots(1)$

$\log_b a = c - 1$

$\Rightarrow \dfrac{\ln a}{\ln b} = c - 1 \quad \ldots(2)$

(1) − (2):

$2 \dfrac{\ln b}{\ln a} - \dfrac{\ln a}{\ln b} = 1$

$2(\ln b)^2 - (\ln a)^2 = \ln a \ln b$

$(\ln a)^2 + \ln b (\ln a) - 2(\ln b)^2 = 0$

Treating this as a quadratic in $\ln a$ and factorising:

$(\ln a - \ln b)(\ln a + 2\ln b) = 0$

$\therefore \ln a = \ln b$ or $\ln a = -2 \ln b$

i.e. $a = b$ or $a = e^{-2\ln b} = e^{\ln b^{-2}} = b^{-2}$

But we are given that $a < b$, so $a \neq b$ and hence $a = b^{-2}$.

9

$\ln x = 4 \log_x e$

$\ln x = 4 \dfrac{\ln e}{\ln x}$

$(\ln x)^2 = 4$

$\ln x = \pm 2$

$x = e^{\pm 2}$

Long questions

1 a As $t \to \infty$, $e^{-0.2t} \to 0$ and so $V \to 42$

When $t = 0$, $V = 42(1 - e^0) = 0$

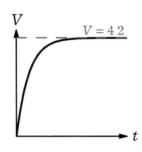

Figure 2ML.1 Graph of $V = 42(1 - e^{-0.2t})$

b When $t = 0$, $V = 42(1 - e^0) = 0 \text{ m s}^{-1}$

c As $t \to \infty$, $e^{-0.2t} \to 0$, $\therefore V \to 42 \text{ m s}^{-1}$

d When $V = 22$,

$22 = 42(1 - e^{-0.2t})$

$1 - e^{-0.2t} = \dfrac{22}{42}$

$e^{-0.2t} = 1 - \dfrac{11}{21} = \dfrac{10}{21}$

$-0.2t = \ln \dfrac{10}{21}$

$0.2t = -\ln \dfrac{10}{21} = \ln \dfrac{21}{10}$

$t = 5 \ln \dfrac{21}{10} = 3.71 \text{ s (3SF)}$

> **COMMENT**
>
> The graph in (a) can be drawn immediately with a GDC; the answers to (b) and (c) can then simply be deduced from the graph without first calculating V when $t = 0$ and the limiting value of V as $t \to \infty$.

2 a i $T = ka^n$

37 000 tigers in 1970, i.e. when $n = 0$, $T = 37\,000$.

$\therefore 37\,000 = ka^0$

$k = 37\,000$

ii 22 000 tigers in 1980, i.e. when $n = 10$, $T = 22\,000$.

$\therefore 22\,000 = 37\,000 a^{10}$

$a^{10} = \dfrac{22}{37}$

$a = \sqrt[10]{\dfrac{22}{37}} = 0.949$ (3SF)

b In 2020, $n = 50$:

$T = 37\,000 \times 0.949^{50} = 2750$

c When $T = 1000$:

$1000 = 37\,000 \times 0.949^n$

$0.949^n = \dfrac{1}{37}$

$\ln 0.949^n = \ln \dfrac{1}{37}$

$n \ln 0.949 = \ln \dfrac{1}{37}$

$n = \dfrac{\ln \dfrac{1}{37}}{\ln 0.949} = 69.5$

so tigers will reach 'near extinction' in 2039.

d Under the initial model, in 2000 ($n = 30$) the number of tigers is

$T = 37\,000 \times 0.949^{30} = 7778$

The new model, $T = kb^m$, has $T = 7778$ when $m = 0$, so $k = 7778$.

Under this model ($T = 7778b^m$), there are 10 000 tigers in 2010, i.e. when $m = 10$, $T = 10\,000$.

$\therefore 10\,000 = 7778 b^{10}$

$b^{10} = \dfrac{10\,000}{7778}$

$b = \sqrt[10]{\dfrac{10\,000}{7778}} = 1.025$

Therefore the new model is $T = 7778 \times 1.025^m$.

e The growth factor each year is 1.025, equivalent to a 2.5% growth rate.

3 a i A 3.5% annual increase $\Rightarrow$ multiplication by 1.035

$\$2000 \times 1.035 = \$2070$

ii $\$2000 \times 1.035^5 = \$2375.37$

b Amount of money after n years is $A = 2000 \times 1.035^n$.

When $A = 4000$,

$4000 = 2000 \times 1.035^n$

$1.035^n = 2$

$n \ln 1.035 = \ln 2$

$n = \dfrac{\ln(2)}{\ln(1.035)} = 20.1$

So it takes approximately 20 years for the amount to double.

c

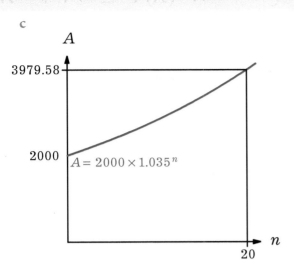

Figure 2ML.3 Graph of $A = 2000 \times 1.035^n$

d Balance in the second account after n years is $B = 3000 \times 1.02^n$.
Balances in the two accounts are the same when

$3000 \times 1.02^n = 2000 \times 1.035^n$

$\dfrac{1.035^n}{1.02^n} = \dfrac{3}{2}$

$\left(\dfrac{1.035}{1.02}\right)^n = 1.5$

$n \ln \dfrac{1.035}{1.02} = \ln 1.5$

$n = \dfrac{\ln(1.5)}{\ln\left(\dfrac{1.035}{1.02}\right)} = 27.8$

So the balances would be the same after approximately 28 years.

e If x dollars is invested in the second account, after n years the balance is $B = x \times 1.02^n$.

After 20 years, the balance in this account is the same as that in first account if

$x \times 1.02^{20} = 2000 \times 1.035^{20}$

$x = 2000 \times \left(\dfrac{1.035}{1.02}\right)^{20} = 2678.14$

So \$2678.14 should be invested in the second account.

3 Algebraic structures

Exercise 3A

2 $6^x - 4 \times 3^x = 0$

$6^x = 4 \times 3^x$

$\dfrac{6^x}{3^x} = 4$

$\left(\dfrac{6}{3}\right)^x = 4$

$2^x = 4$

$x = 2$

3 $2 \times 5^x - 7 \times 10^x = 0$

$\dfrac{10^x}{5^x} = \dfrac{2}{7}$

$\left(\dfrac{10}{5}\right)^x = \dfrac{2}{7}$

$2^x = \dfrac{2}{7}$

$x \log 2 = \log \dfrac{2}{7}$

$x = \dfrac{\log\left(\dfrac{2}{7}\right)}{\log 2}$

$= \dfrac{\log 2 - \log 7}{\log 2}$

4 $(3x-1)^{x^2-4} = 1$

$\therefore \begin{cases} 3x-1 = 1 \\ \text{or } 3x-1 = -1 \text{ and } x^2-4 \text{ is even} \\ \text{or } x^2-4 = 0 \text{ and } 3x-1 \neq 0 \end{cases}$

i.e. $x = \dfrac{2}{3}$ or $x = 0$ or $x = \pm 2$

Exercise 3B

3 a $e^{2x} - 9e^x + 20 = 0$

$(e^x)^2 - 9e^x + 20 = 0$

Let $y = e^x$; then

$y^2 - 9y + 20 = 0$

$(y-4)(y-5) = 0$

$y = 4$ or $y = 5$

$\therefore e^x = 4$ or $e^x = 5$

$x = \ln 4$ or $x = \ln 5$

b $4^x - 7 \times 2^x + 12 = 0$

$(2^x)^2 - 7(2^x) + 12 = 0$

Let $y = 2^x$; then

$y^2 - 7y + 12 = 0$

$(y-3)(y-4) = 0$

$y = 3$ or $y = 4$

$\therefore 2^x = 3$ or $2^x = 4$

$x = \dfrac{\log 3}{\log 2}$ or $x = 2$

c $(\log_3 x)^2 - 3\log_3 x + 2 = 0$

Let $y = \log_3 x$; then

$y^2 - 3y + 2 = 0$

$(y-1)(y-2) = 0$

$y = 1$ or $y = 2$

$\therefore \log_3 x = 1$ or $\log_3 x = 2$

$x = 3^1 = 3$ or $x = 3^2 = 9$

4 $9(1+9^{x-1})=10\times 3^x$

$9^x+9-10\times 3^x=0$

$(3^x)^2-10(3^x)+9=0$

Let $u=3^x$:

$u^2-10u+9=0$

$(u-1)(u-9)=0$

$u=1$ or $u=9$

$\therefore 3^x=1$ or $3^x=9$

$x=0$ or $x=2$

5 $a^x=-\dfrac{5}{a^x}+6$

$a^x-6+\dfrac{5}{a^x}=0$

$(a^x)^2-6a^x+5=0$

Let $u=a^x$:

$u^2-6u+5=0$

$(u-1)(u-5)=0$

$u=1$ or $u=5$

$\therefore a^x=1$ or $a^x=5$

$x=0$ or $x=\dfrac{\log 5}{\log a}$

6 $\log_2 x=6-5\log_x 2$

$\log_2 x=6-5\dfrac{\log_2 2}{\log_2 x}$

$(\log_2 x)^2-6\log_2 x+5=0$

Let $u=\log_2 x$:

$u^2-6u+5=0$

$(u-1)(u-5)=0$

$u=1$ or $u=5$

$\therefore \log_2 x=1$ or $\log_2 x=5$

$x=2^1=2$ or $x=2^5=32$

> **COMMENT**
>
> Instead of changing $\log_x$ into $\log_2$, the opposite could have been done, or both $\log_x$ and $\log_2$ could have been changed to log (base 10).

Exercise 3C

> **COMMENT**
>
> The worked solutions for this exercise set contain text illustrating the necessary thought process to ensure a full and correct answer; this level of detailed narrative would not be required in an examination solution.

2 $y=x\ln x$

$\ln x$ ceases to be defined at $x=0$; as $x\to 0$, $x\ln x\to 0$, so there is no vertical asymptote, but there is an empty circle at the origin and no graph for $x<0$.

As x gets large, both x and $\ln x$ continue to increase, so there is no horizontal asymptote.

$x\ln x=0$ when $x=0$, or $\ln x=0$. The first root has already been eliminated, so the only true root is when $\ln x=0$: $x=1$.

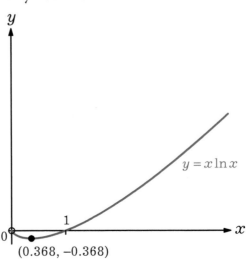

(0.368, −0.368)

Figure 3C.2

3 Algebraic structures

3 $y = \dfrac{e^x}{\ln x}$

$\ln x = 0$ at $x = 1$, so there is a vertical asymptote at $x = 1$.

$\ln x$ ceases to be defined at $x = 0$, and as $x \to 0$ from above, $\ln x \to -\infty$, so the graph terminates with an empty circle at the origin.

For large x, e^x increases more rapidly than $\ln x$, so their ratio increases and there is no horizontal asymptote.

$y = 0$ when $e^x = 0$, which has no solutions, so there are no roots.

The exact value of the minimum is best found using a GDC, but from the above we can be confident that there must be a single minimum at some $x > 1$.

For large positive x, e^x grows more rapidly than the quartic numerator, so $y = 0$ will be an asymptote.

For large negative x, e^x approaches zero, so the ratio will grow without limit.

Maximum and minimum points are best found using a GDC, but we can be confident that, since there are no vertical asymptotes, there must be a minimum between -3 and 0, another minimum between 0 and 3, and then a maximum at some point with $x > 3$, in addition to the maximum at 0.

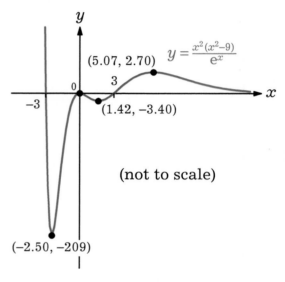

(not to scale)

Figure 3C.4

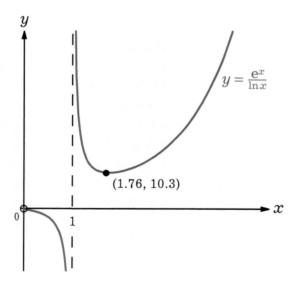

Figure 3C.3

4 $y = \dfrac{x^2(x^2 - 9)}{e^x}$

$e^x \neq 0$, so there are no vertical asymptotes.

$y = 0$ when $x^2(x^2 - 9) = 0$, which occurs when $x = 0$ or $x = \pm 3$, so there are three roots, with the root at 0 being a twice-repeated root (so that the graph will touch the axis rather than crossing it).

Exercise 3D

2 $x \ln x = 3 - x^2$

The graphs of $y = x \ln x$ and $y = 3 - x^2$ intersect at one point.

From GDC: $x = 1.53$ (3SF)

3 For $\ln x = kx$ to have exactly one solution, the graph of $y = kx$ must be tangent to the graph of $y = \ln x$.

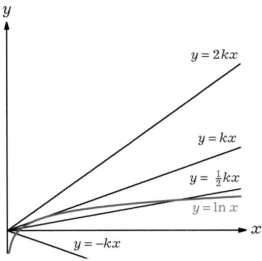

Figure 3D.3 Graphs of $y = kx$ and $y = \ln x$

a $\ln x^2 = 2 \ln x$, so $\ln x^2 = kx$ is equivalent to $\ln x = \dfrac{k}{2} x$, which will have two solutions since the line $y = \dfrac{k}{2} x$ has a smaller gradient than $y = kx$.

b $\ln \left(\dfrac{1}{x} \right) = \ln x^{-1} = -\ln x$, so $\ln \left(\dfrac{1}{x} \right) = kx$ is equivalent to $\ln x = -kx$, which will have one solution, since there will be an intersection in the lower right quadrant.

c $\ln \sqrt{x} = \ln x^{\frac{1}{2}} = \dfrac{1}{2} \ln x$, so $\ln \sqrt{x} = kx$ is equivalent to $\ln x = 2kx$, which will have no solutions since the line $y = 2kx$ has a greater gradient than $y = kx$.

4 Considering the graph of $y = \sin x$, $\sin x = kx$ will have one solution at $x = 0$ and any solution $x = a$ will be matched by a solution at $x = -a$.

For exactly seven solutions, there must be three positive solutions; this can happen when the line crosses the curve once in each of the intervals $[90, 180[$, $[360, 450[$ and $[450, 540[$ but not in $[720, 810]$.

An easy way to accomplish this would be to have the line and the curve intersect at $(450, 1)$, so that $k = \dfrac{1}{450} = 0.00222$ (3SF).

> **COMMENT**
>
> It is beyond the scope of this question to determine the full set of possible values for k; the question only requires one example value, so select one that is easy to establish. It is left to the ambitious student, using techniques from Chapter 12, to show that the boundary values for the complete solution set for seven intersection points satisfy the equation
>
> $x = \dfrac{180}{\pi} \tan \dfrac{\pi x}{180}$, giving rise to the solution set $k \in \,]0.00124, 0.00224[\cup \{-0.00159\}$, correct to 3SF.

Mixed examination practice 3

1 a $y = 2^x$: axis intercept at $(0, 1)$, exponential shape.

$y = 1 - x^2$: axis intercepts at $(0, 1)$ and $(\pm 1, 0)$, negative quadratic.

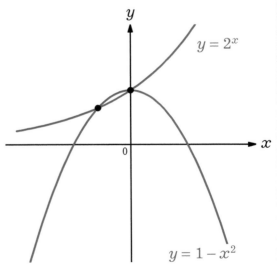

Figure 3M.1

b Two intersection points $\Rightarrow$ two solutions of $2^x = 1 - x^2$.

2 Sketching the graph on GDC:

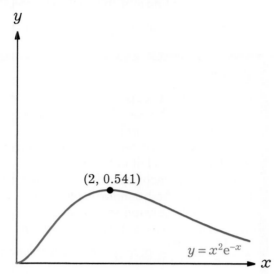

Figure 3M.2

Maximum value of y is 0.541 (3SF)

3 $e^x \ln x = 3e^x$

$e^x (\ln x - 3) = 0$

$\therefore \ln x = 3$ (no solutions to $e^x = 0$)

$x = e^3$

4 Vertical asymptotes where the denominator equals zero:

$ax + b = 0$ or $x - c = 0$

$x = -\dfrac{b}{a}$ and $x = c$

5 a

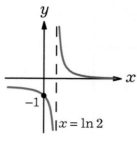

Figure 3M.5 Graph of $y = \dfrac{1}{e^x - 2}$

b Vertical asymptote where the denominator equals zero:

$e^x - 2 = 0$

$e^x = 2$

$x = \ln 2$

6 $x^4 + 36 = 13x^2$

$(x^2)^2 - 13x^2 + 36 = 0$

Let $u = x^2$:

$u^2 - 13u + 36 = 0$

$(u-9)(u-4) = 0$

$u = 9$ or $u = 4$

$\therefore x^2 = 4$ or 9

$x = \pm 2, \pm 3$

7 $x \ln x + 4 \ln x = 0$

$(x+4) \ln x = 0$

$x = -4$ or $\ln x = 0$

$\therefore x = 1$

(as $\ln x$ has no real value for $x = -4$)

> **COMMENT**
> Always check the validity of algebraic solutions, especially in functions with restricted domains, such as rational functions and those containing logarithms.

4 The theory of functions

Exercise 4B

4 $f(x) = \sqrt{\ln(x-4)}$

Square root can have only non-negative values in its domain, so require $\ln(x-4) \geq 0$:

$x - 4 \geq e^0$

$x \geq 5$

Domain of $f(x)$ is $x \geq 5$

5 $f(x) = \dfrac{4^{\sqrt{x-1}}}{x+2} - \dfrac{1}{(x-3)(x-2)} + x^2 + 1$

Cannot have division by zero, so $x \neq -2, 2, 3$

Square root can only have non-negative values in its domain, so require $x - 1 \geq 0$, i.e. $x \geq 1$

Domain of $f(x)$ is $x \geq 1, x \neq 2, x \neq 3$

> **COMMENT**
> Note that the restriction $x \neq -2$ is not needed in the final answer as it is already covered by the restriction $x \geq 1$.

6 $g(x) = \ln(x^2 + 3x + 2)$

$\ln x$ can have only positive values in its domain, so require $x^2 + 3x + 2 > 0$:

$x^2 + 3x + 2 > 0$

$(x+1)(x+2) > 0$

$x < -2$ or $x > -1$

Domain of $g(x)$ is $x < -2$ or $x > -1$

> **COMMENT**
> It may be helpful to draw a graph of $y = x^2 + 3x + 2$ to solve the quadratic inequality $x^2 + 3x + 2 > 0$.

7 $f(x) = \sqrt{\dfrac{8x-4}{x-12}}$

Cannot have division by zero $\Rightarrow x \neq 12$

Square root can have only non-negative values in its domain, so require either $8x - 4 \geq 0$ and $x - 12 > 0$ or $8x - 4 \leq 0$ and $x - 12 < 0$.

$8x - 4 \geq 0$ and $x - 12 > 0$

$\Rightarrow x \geq \dfrac{1}{2}$ and $x > 12$

$\therefore x > 12$

$8x - 4 \leq 0$ and $x - 12 < 0$

$\Rightarrow x \leq \dfrac{1}{2}$ and $x < 12$

$\therefore x \leq \dfrac{1}{2}$

Domain of $f(x)$ is $x \leq \dfrac{1}{2}$ or $x > 12$

8 $f(x) = \sqrt{x-a} + \ln(b-x)$

a Square root can have only non-negative values in its domain, so require $x \geq a$

$\ln x$ can have only positive values in its domain, so require $x < b$

i $a < b \Rightarrow$ domain is $a \leq x < b$

ii $a > b \Rightarrow$ function has empty domain

b $f(a) = \begin{cases} \sqrt{a-a} + \ln(b-a) & \text{if } a < b \\ \text{undefined} & \text{if } a \geq b \end{cases}$

$= \begin{cases} \ln(b-a) & \text{if } a < b \\ \text{undefined} & \text{if } a \geq b \end{cases}$

Exercise 4C

3 $fg(x) = (3x+2)^2 + 1$
$= 9x^2 + 12x + 5$

$gf(x) = 3(x^2+1) + 2$
$= 3x^2 + 5$

$fg(x) = gf(x)$
$9x^2 + 12x + 5 = 3x^2 + 5$
$6x^2 + 12x = 0$
$6x(x+2) = 0$
$x = 0$ or $x = -2$

4 $gf(x) = 0$
$\dfrac{3x+1}{(3x+1)^2 + 25} = 0$
$3x + 1 = 0$
$\therefore x = -\dfrac{1}{3}$

5 a Horizontal asymptote is $y = 2$, so the range is $y \neq 2$

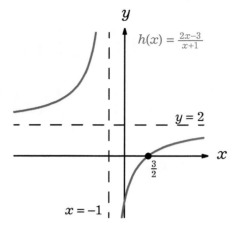

Figure 4C.5.1 Graph of $h(x) = \dfrac{2x-3}{x+1}$

b $h(x) = 0$
$\dfrac{2x-3}{x+1} = 0$
$2x - 3 = 0$
$x = \dfrac{3}{2}$

c To define $g \circ h$, the range of h must be a subset of the domain of g. Domain of $g(x)$ is $x \geq 0$, so need to restrict the domain of h so that the range of $h(x)$ is $y \geq 0$. (Without restriction, the domain of h is $x \neq -1$.)

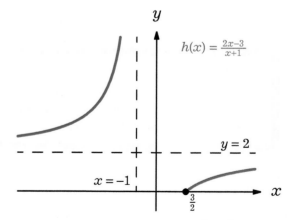

Figure 4C.5.2 Graph of $h(x) = \dfrac{2x-3}{x+1}$ with domain restricted so that the range is $y \geq 0$

Hence the domain, D, of $g \circ h(x)$ is $x < -1$ or $x \geq \dfrac{3}{2}$

Range of h over domain D is $y \geq 0$, $y \neq 2$

Range of g over domain $x \geq 0$, $x \neq 2$ is $y \geq 0$, $y \neq \sqrt{2}$

$\therefore$ range of $g \circ h$ over domain D is $y \geq 0$, $y \neq \sqrt{2}$

6 a $fg(x) = 2x + 3$
$[g(x)]^3 = 2x + 3$
$\Rightarrow g(x) = \sqrt[3]{2x + 3}$

b $gf(x) = 2x + 3$
$g(x^3) = 2x + 3$
$\Rightarrow g(x) = 2\sqrt[3]{x} + 3$

7 a $fg(x) = \sqrt{(3x+4)^2 - 2(3x+4)}$

$f \circ g$ is undefined for $x \in \,]a,b[$, so require $(3x+4)^2 - 2(3x+4) < 0$ (since square root is undefined for negative values).

$(3x+4)^2 - 2(3x+4) < 0$
$(3x+4)[(3x+4) - 2] < 0$
$(3x+4)(3x+2) < 0$
$x \in \left]-\dfrac{4}{3}, -\dfrac{2}{3}\right[$

$\therefore a = -\dfrac{4}{3},\ b = -\dfrac{2}{3}$

b Over the domain $x \notin \,]a, b[$, $(3x+4)^2 - 2(3x+4)$ takes all non-negative values and so $fg(x) = \sqrt{(3x+4)^2 - 2(3x+4)}$ takes all non-negative values, i.e. the range of $f \circ g$ is $y \geq 0$.

8 a The range of f is $y > 2$; this lies within the domain of g, so $g \circ f$ is a valid composition.

The range of g is $y \geq 0$; values from $[0, 3]$ lie within the range of g but not within the domain of f, so $f \circ g$ is not a valid composition for the full domain of g.

b For $f \circ g$ to be defined, we require the range of g to be limited to $]3, \infty[$, so restrict the domain to $x \notin \left[-\sqrt{3}, \sqrt{3}\right]$.

9 By observation,

$g\left(\dfrac{x}{2} - 3\right) = 2\left(\dfrac{x}{2} - 3\right) + 5$
$= x - 6 + 5$
$= x - 1$

$\therefore f(x-1) = fg\left(\dfrac{x}{2} - 3\right)$

$= \dfrac{\dfrac{x}{2} - 3 + 2}{3}$

$= \dfrac{x}{6} - \dfrac{1}{3}$

Alternatively, given that $fg(x) = \dfrac{x+2}{3}$

and $g(x) = 2x + 5$, we have $f(2x+5) = \dfrac{x+2}{3}$.

Let $2x + 5 = u - 1$, so that $x = \dfrac{u-6}{2} = \dfrac{u}{2} - 3$.
Then

$f(u-1) = \dfrac{\dfrac{u}{2} - 3 + 2}{3} = \dfrac{u}{6} - \dfrac{1}{3}$

$\therefore f(x-1) = \dfrac{x}{6} - \dfrac{1}{3}$

Exercise 4D

3 a $ff(2) = f(0) = -1$

b $f(1) = 3$ so $f^{-1}(3) = 1$

4 $y = \sqrt{3-2x}$

$3 - 2x = y^2$

$\Rightarrow x = \dfrac{3-y^2}{2}$

$\therefore f^{-1}(x) = \dfrac{3-x^2}{2}$

Hence $f^{-1}(7) = \dfrac{3-7^2}{2} = -23$

5 $y = 3e^{2x}$

$e^{2x} = \dfrac{y}{3}$

$2x = \ln \dfrac{y}{3}$

$\Rightarrow x = \dfrac{1}{2}\ln\left(\dfrac{y}{3}\right) = \ln\sqrt{\dfrac{y}{3}}$

$\therefore f^{-1}(x) = \ln\sqrt{\dfrac{x}{3}}$

The range of f is $y > 0$, so the domain of f^{-1} is $x > 0$.

6 $fg(x) = 2(x^3) + 3$

$y = 2x^3 + 3$

$x^3 = \dfrac{y-3}{2}$

$\Rightarrow x = \sqrt[3]{\dfrac{y-3}{2}}$

$\therefore (fg)^{-1}(x) = \sqrt[3]{\dfrac{x-3}{2}}$

7 a To find the inverse function of $f(x) = e^{2x}$:

$y = e^{2x}$

$2x = \ln y$

$\Rightarrow x = \dfrac{1}{2}\ln y$

$\therefore f^{-1}(x) = \dfrac{1}{2}\ln x = \ln\sqrt{x}$

To find the inverse function of $g(x) = x + 1$:

$y = x + 1$

$\Rightarrow x = y - 1$

$\therefore g^{-1}(x) = x - 1$

So

$f^{-1}(3) \times g^{-1}(3) = \left(\ln\sqrt{3}\right) \times (3-1)$

$= 2\ln\sqrt{3}$

$= \ln 3$

b $(fg)^{-1}(x) = g^{-1}f^{-1}(x) = \ln\sqrt{x} - 1$

$\therefore (fg)^{-1}(3) = \ln\sqrt{3} - 1$

8 $f(x) = \sqrt{x} \Rightarrow f^{-1}(x) = x^2$ for $x \geq 0$

$(f^{-1} \circ g)(x) = (2x)^2$

$\therefore (f^{-1}g)(x) = 0.25$

$4x^2 = 0.25$

$x^2 = \dfrac{1}{16}$

$x = \dfrac{1}{4}$

> **COMMENT**
>
> $x = -\dfrac{1}{4}$ is not a valid solution here, as it is not in the domain of $(f^{-1}g)(x)$. This is because $g\left(-\dfrac{1}{4}\right) = -\dfrac{1}{2}$ and $-\dfrac{1}{2}$ is not in the domain of f^{-1}. (The domain of f^{-1} is the range of f, and the latter consists of non-negative values only.)

9 $y = \dfrac{x^2-4}{x^2+9}$

$(x^2+9)y = x^2-4$

$x^2 y + 9y = x^2 - 4$

$x^2 y - x^2 = -4 - 9y$

$x^2(y-1) = -4 - 9y$

$x^2 = -\dfrac{4+9y}{y-1}$

$= \dfrac{4+9y}{1-y}$

$\therefore x = -\sqrt{\dfrac{4+9y}{1-y}}$ (as domain of f is $x \le 0$)

Hence $f^{-1}(x) = -\sqrt{\dfrac{4+9x}{1-x}}$

The graph of f has a horizontal asymptote at $y = 1$ (as $x \to -\infty$) and is decreasing for all $x \le 0$, so the range of f is $y \in \left[-\dfrac{4}{9}, 1\right[$.

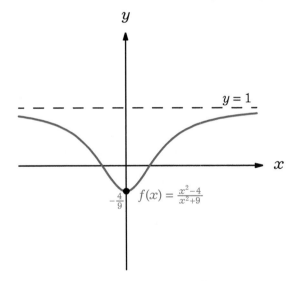

Figure 4D.9 Graph of $f(x) = \dfrac{x^2-4}{x^2+9}$ for real x

Hence the domain of f^{-1} is $x \in \left[-\dfrac{4}{9}, 1\right[$.

10 a $y = \ln(x-1) + \ln 3$

$= \ln(3x-3)$

$3x - 3 = e^y$

$\Rightarrow x = \dfrac{e^y + 3}{3}$

$= \dfrac{e^y}{3} + 1$

$\therefore f^{-1}(x) = \dfrac{e^x}{3} + 1$

The range of f is $\mathbb{R}$, so the domain of f^{-1} is also $\mathbb{R}$.

b $f(x) = \ln(x-1) + \ln(3)$

$= \ln[3(x-1)]$

$= \ln(3x-3)$

$\therefore gf(x) = e^{\ln(3x-3)}$

$= 3x - 3$

11 a Finding the inverse of $f(x) = \dfrac{1}{x}$:

$y = \dfrac{1}{x}$

$\Rightarrow x = \dfrac{1}{y}$

$\therefore f^{-1}(x) = \dfrac{1}{x}$

So $f^{-1}(x) = \dfrac{1}{x} = f(x)$, i.e. f is self-inverse.

b Finding the inverse of $g(x) = \dfrac{3x-5}{x+k}$:

$y = \dfrac{3x-5}{x+k}$

$y(x+k) = 3x-5$

$xy + ky = 3x - 5$

$3x - xy = 5 + ky$

$x(3-y) = 5 + ky$

$\Rightarrow x = \dfrac{5+ky}{3-y}$

$\therefore g^{-1}(x) = \dfrac{5+kx}{3-x}$

4 The theory of functions   31

Require that $g(x) = g^{-1}(x)$ for all x:

$$\frac{3x-5}{x+k} = \frac{5+kx}{3-x}$$

$$\Rightarrow \frac{3x-5}{x+k} = \frac{-kx-5}{x-3}$$

Comparing these, it is evident that $k = -3$.

> **COMMENT**
>
> If it is difficult to see that multiplying the numerator and denominator as above enables a straightforward comparison to determine k, then the following (more lengthy!) process can be undertaken instead:
>
> $$\frac{3x-5}{x+k} = \frac{5+kx}{3-x}$$
> $$(3x-5)(3-x) = (5+kx)(x+k)$$
> $$9x - 3x^2 - 15 + 5x = 5x + 5k + kx^2 + k^2x$$
> $$-3x^2 + 14x - 15 = kx^2 + (k^2+5)x + 5k$$
>
> Comparing coefficients of the two sides:
>
> x^2: $-3 = k$
> x^1: $14 = k^2 + 5$
> x^0: $-15 = 5k$
>
> These three equations consistently give the unique solution $k = -3$.

6 $f(x) = \dfrac{1}{x+3}$

a Cannot have division by zero, so domain is $x \neq -3$

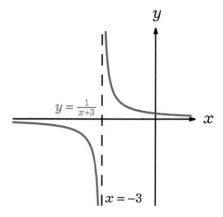

Figure 4E.6

Range is $y \neq 0$

b $y = \dfrac{1}{x+3}$

$y(x+3) = 1$

$xy = 1 - 3y$

$\Rightarrow x = \dfrac{1-3y}{y}$

$\therefore f^{-1}(x) = \dfrac{1-3x}{x}$

7 a Horizontal asymptote: $y = 0$
Vertical asymptote: $x = 0$

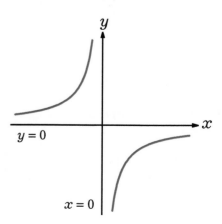

Figure 4E.7

Exercise 4E

5 $y = \dfrac{3x-1}{4-5x}$

Vertical asymptote where denominator equals zero: $x = \dfrac{4}{5}$

Horizontal asymptote as $x \to \pm\infty$: $y = -\dfrac{3}{5}$

b $y = -\dfrac{3}{x}$

$\Rightarrow x = -\dfrac{3}{y}$

$\therefore f^{-1}(x) = -\dfrac{3}{x}, \quad x \neq 0$

8 $y = \dfrac{3x-1}{x-5}$

Vertical asymptote where denominator equals zero: $x = 5$

Horizontal asymptote as x gets large: $y = 3$

Axis intercepts: $\left(0, \dfrac{1}{5}\right)$ and $\left(\dfrac{1}{3}, 0\right)$

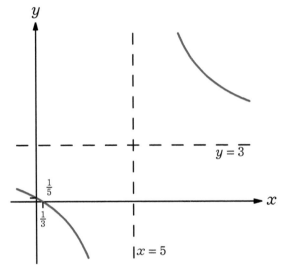

Figure 4E.8

9 $f(x) = \dfrac{ax+3}{2x-8}, \quad x \neq 4$

a Horizontal asymptote as x gets large: $y = \dfrac{a}{2}$

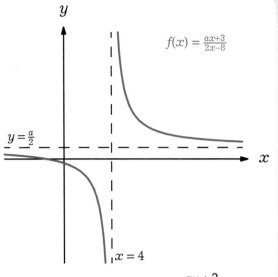

Figure 4E.9 Graph of $f(x) = \dfrac{ax+3}{2x-8}$ for positive a

Range of f is $y \in \mathbb{R}, \quad y \neq \dfrac{a}{2}$

b $y = \dfrac{ax+3}{2x-8}$

$y(2x-8) = ax+3$

$2xy - 8y = ax + 3$

$2xy - ax = 8y + 3$

$x(2y - a) = 8y + 3$

$\Rightarrow x = \dfrac{8y+3}{2y-a}$

$\therefore f^{-1}(x) = \dfrac{8x+3}{2x-a}, \quad x \neq \dfrac{a}{2}$

(The domain of f^{-1} is the range of f.)

c For f to be self-inverse, require that $f^{-1}(x) = f(x)$ for all x.

The vertical asymptote of f is $x = 4$; this must be the same as the vertical asymptote of f^{-1}, which is $x = \dfrac{a}{2}$:

$\dfrac{a}{2} = 4 \Rightarrow a = 8$

Mixed examination practice 4

Short questions

1 $f(x) = x^2 + 1$
$$\therefore f(2x-1) = (2x-1)^2 + 1$$
$$= 4x^2 - 4x + 2$$

2 $f(x) = x+2, \quad g(x) = x^2$
$fg(x) = gf(x)$
$x^2 + 2 = (x+2)^2$
$x^2 + 2 = x^2 + 4x + 4$
$4x = -2$
$x = -\dfrac{1}{2}$

3 $y = e^{2x}$
$2x = \ln y$
$\Rightarrow x = \dfrac{1}{2}\ln y$
$\therefore f^{-1}(x) = \dfrac{1}{2}\ln x$

$f^{-1}(3) = \dfrac{1}{2}\ln 3 = 0.549$ (3SF)

4 a Vertical asymptote where denominator equals zero: $x = 5$

Horizontal asymptote for large x:
$y = \dfrac{4}{-1} = -4$

b $y = \dfrac{4x-3}{5-x}$
$(5-x)y = 4x-3$
$5y - xy = 4x - 3$
$4x + xy = 5y + 3$
$x(4+y) = 5y + 3$
$\Rightarrow x = \dfrac{5y+3}{y+4}$
$\therefore f^{-1}(x) = \dfrac{5x+3}{x+4}$

5 a $y = \log_3(x+3)$
$3^y = x + 3$
$\Rightarrow x = 3^y - 3$
$\therefore f^{-1}(x) = 3^x - 3$

(Range of f is $\mathbb{R}$, so domain of f^{-1} is also $\mathbb{R}$.)

b $y = 3e^{x^3 - 1}$
$\dfrac{y}{3} = e^{x^3 - 1}$
$\ln\left(\dfrac{y}{3}\right) = x^3 - 1$
$\Rightarrow x = \left(1 + \ln\left(\dfrac{y}{3}\right)\right)^{\frac{1}{3}}$
$\therefore f^{-1}(x) = \left(1 + \ln\left(\dfrac{x}{3}\right)\right)^{\frac{1}{3}}$

(Range of f is $y > 0$, so domain of f^{-1} is $x > 0$.)

6 a Reflection of $f(x)$ in the line $y = x$ gives the graph of $f^{-1}(x)$, so C is $y = \log_2 x$.

b C cuts the x-axis where $y = 0$:
$\log_2 x = 0$
$\Rightarrow x = 2^0 = 1$
i.e. intersection at (1, 0).

7 a $f(x) = x^2 - 6x + 10$
$= (x-3)^2 - 9 + 10$
$= (x-3)^2 + 1$

b $y = (x-3)^2 + 1$
$\Rightarrow x - 3 = \sqrt{y-1}$
(the positive square root is needed as $x \geq 3$)
$\Rightarrow x = 3 + \sqrt{y-1}$
$\therefore f^{-1}(x) = 3 + \sqrt{x-1}$

c   The minimum point of f is $(3, 1)$, so the range of f is $y \geq 1$ and hence the domain of f^{-1} is $x \geq 1$.

8 a   Horizontal asymptote is $y = -1$, so the range is $y \neq -1$.

b   Vertical asymptote: $x = -1$

Axis intercepts: $(3, 0)$ and $(0, 3)$

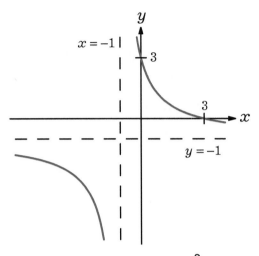

Figure 4MS.8   Graph of $f(x) = \dfrac{3-x}{x+1}$

c $y = \dfrac{3-x}{x+1}$

$(x+1)y = 3-x$

$xy + y = 3 - x$

$xy + x = 3 - y$

$x(y+1) = 3 - y$

$\Rightarrow x = \dfrac{3-y}{y+1}$

$\therefore f^{-1}(x) = \dfrac{-x+3}{x+1}$

Domain and range of f are $x \neq -1$ and $y \neq -1$, so domain and range of f^{-1} are $x \neq -1$ and $y \neq -1$.

9 a $h(x) = x^2 - 6x + 2$

$= (x-3)^2 - 9 + 2$

$= (x-3)^2 - 7$

b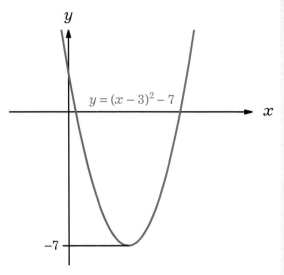

Figure 4MS.9

The domain of h is $x > 3$, so the range of h is $y > -7$

c $y = (x-3)^2 - 7$

$\Rightarrow x - 3 = \sqrt{y+7}$

(the positive square root is needed as $x > 3$)

$\Rightarrow x = 3 + \sqrt{y+7}$

$\therefore h^{-1}(x) = 3 + \sqrt{x+7}$, $x > -7$

10 a $f(x) = \sqrt{x-2}$, $g(x) = x^2 + x$

$\therefore fg(x) = \sqrt{x^2 + x - 2}$

$f \circ g$ is undefined when $x^2 + x - 2 < 0$ (since the square root is undefined for negative values):

$x^2 + x - 2 < 0$

$(x+2)(x-1) < 0$

$x \in \,]-2, 1[$

$\therefore a = -2, b = 1$

4 The theory of functions   35

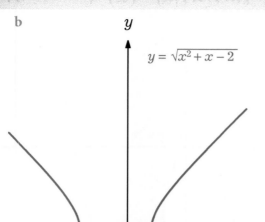

Figure 4MS.10 Graph of $fg(x) = \sqrt{x^2 + x - 2}$

Range of g on the domain $\mathbb{R} - \,]-2, 1[$ is $y \geq 2$

Range of f on the domain $x \geq 2$ is $y \geq 0$

$\therefore$ range of $f \circ g$ on the domain $\mathbb{R} - \,]-2, 1[$ is $y \geq 0$

Long questions

1 a $f(3) = 3^2 + 1 = 10$

b $gf(x) = g(x^2 + 1)$
$= 5 - (x^2 + 1)$
$= 4 - x^2$

c The graphs of a function and its inverse are reflections of each other in the line $y = x$.

d i $y = x^2 + 1$
$x^2 = y - 1$
$\Rightarrow x = \sqrt{y - 1}$
(the positive square root is needed as $x > 3$)
$\therefore f^{-1}(x) = \sqrt{x - 1}$

ii Domain of f is $x \geq 3$, so range of f^{-1} is $y \geq 3$.

iii Range of f is $y \geq 10$, so domain of f^{-1} is $x \geq 10$.

e $f(x) = g(3x)$
$x^2 + 1 = 5 - 3x$
$x^2 + 3x - 4 = 0$
$(x + 4)(x - 1) = 0$
$x = -4$ or $x = 1$

However, the domain of f is $x > 3$ so there are no solutions to this equation.

2 a i $f(7) = 2 \times 7 + 1 = 15$

ii Range of f is $\mathbb{R}$

iii $f(z) = 2z + 1$

iv $fg(x) = f\left(\dfrac{x+3}{x-1}\right)$
$= 2\left(\dfrac{x+3}{x-1}\right) + 1$
$= \dfrac{2x+6}{x-1} + \dfrac{x-1}{x-1}$
$= \dfrac{3x+5}{x-1}$

v $ff(x) = 2(2x + 1) + 1$
$= 4x + 3$

b The value $f(0) = 1$ is in the range of f but not in the domain of g, so $gf(0)$ is not defined.

c i $y = \dfrac{x+3}{x-1}$
$(x - 1)y = x + 3$
$xy - y = x + 3$
$xy - x = y + 3$
$x(y - 1) = y + 3$
$\Rightarrow x = \dfrac{y+3}{y-1}$
$\therefore g^{-1}(x) = \dfrac{x+3}{x-1}$

36 Mixed examination practice 4

> **COMMENT**
> Note that $g(x)$ is self-inverse.

ii The graphs of a function and its inverse are reflections of each other in the line $y = x$.

iii $g(x)$ is self-inverse, so the domain and range of $g(x)$ and the domain and range of $g^{-1}(x)$ must all be the same.

∴ domain of $g^{-1}(x)$ is $x \neq 1$.

iv Range of $g^{-1}(x)$ is $y \neq 1$.

3 a $f(x) = x^2 + 4x + 9$
$= (x+2)^2 - 4 + 9$
$= (x+2)^2 + 5$

b $f(x)$ is a positive quadratic with vertex at $(-2, 5)$ and y-intercept at $(0, 9)$.

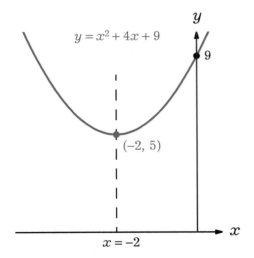

Figure 4ML.3 Graph of $f(x) = x^2 + 4x + 9$

c Range of $f(x)$ is $y \geq 5$.

Range of $g(x)$ is $y > 0$.

d Note that $h(x) = fg(x)$.

Range of f on a domain $x > 0$ is $y > 9$.

∴ range of $h(x)$ is $y > 9$.

4 a $(2x+3)(4-y) = 12$
$8x + 12 - y(2x+3) = 12$
$y(2x+3) = 8x$
$y = \dfrac{8x}{2x+3}$

b Vertical asymptote where denominator equals zero: $x = -\dfrac{3}{2}$

Horizontal asymptote: $y = \dfrac{8}{2} = 4$

Single axis intercept at $(0, 0)$

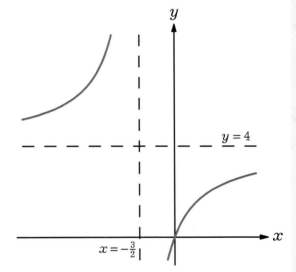

Figure 4ML.4 Graph of $y = \dfrac{8x}{2x+3}$

c $hg(x) = h(2x+k)$
$= \dfrac{8(2x+k)}{2(2x+k)+3}$
$= \dfrac{16x+8k}{4x+2k+3}$

4 The theory of functions 37

d Vertical asymptote where denominator equals zero: $x = -\dfrac{2k+3}{4}$

Horizontal asymptote: $y = \dfrac{16}{4} = 4$

e For $k = -\dfrac{19}{2}$,

$$hg(x) = \dfrac{16x + 8\left(-\dfrac{19}{2}\right)}{4x + 2\left(-\dfrac{19}{2}\right) + 3} = \dfrac{16x - 76}{4x - 16}$$

Finding the inverse:

$$y = \dfrac{16x - 76}{4x - 16}$$

$(4x - 16)y = 16x - 76$

$4xy - 16y = 16x - 76$

$4xy - 16x = 16y - 76$

$(4y - 16)x = 16y - 76$

$\Rightarrow x = \dfrac{16y - 76}{4y - 16}$

$\therefore (hg)^{-1}(x) = \dfrac{16x - 76}{4x - 16} = hg(x)$

i.e. $hg(x)$ is self-inverse.

5 Transformations of graphs

Exercise 5D

1 a $y = p(f(x) + c)$

   b $y = f\left(\dfrac{1}{q}(x+d)\right)$

> **COMMENT**
>
> Notice that the vertical transformations are exactly as you would expect, in terms of both the operations (addition and multiplication) and the order in which they are applied. The horizontal transformations, however, are exactly the opposite of what you might expect in both regards: addition of d and then division by q (rather than multiplication by q and then subtraction of d).

8 $f(x) = 2^x + x$

Vertical stretch with scale factor 8:
multiply through by $8 \Rightarrow$
$f_2(x) = 8(2^x + x)$

Translation by $\begin{pmatrix}1\\4\end{pmatrix}$: replace x with $(x-1)$,

add $4 \Rightarrow f_3(x) = 8(2^{x-1} + x - 1) + 4$

Horizontal stretch with scale factor $\dfrac{1}{2}$:
replace x with $2x \Rightarrow$

$f_4(x) = 8(2^{2x-1} + 2x - 1) + 4$
$ = 2^3 \times 2^{2x-1} + 16x - 4$
$ = 2^{2x+2} + 16x - 4$
$ = (2^2)^{x+1} + 16x - 4$
$ = 4^{x+1} + 16x - 4$

So $h(x) = 4^{x+1} + 16x - 4$

9 a Graph of $y = \ln x$:
vertical asymptote $x = 0$; intercept $(1, 0)$

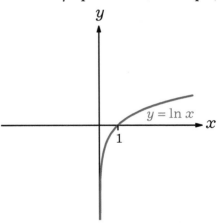

Figure 5D.9.1

   b Graph of $y = 3\ln(x+2)$ is obtained from the graph of $y = \ln x$ by:

translation $\begin{pmatrix}-2\\0\end{pmatrix}$ and vertical stretch with scale factor 3

$\Rightarrow$ vertical asymptote $x = -2$; intercept $(-1, 0)$

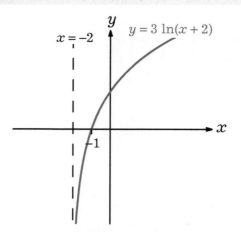

Figure 5D.9.2

c Graph of $y = \ln(2x-1)$ is obtained from the graph of $y = \ln x$ by:

translation $\begin{pmatrix} 1 \\ 0 \end{pmatrix}$ and horizontal stretch with scale factor $\frac{1}{2}$

$\Rightarrow$ vertical asymptote $x = \frac{1}{2}$, intercept $(1, 0)$.

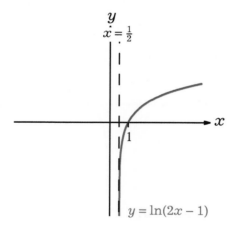

Figure 5D.9.3

10 a $f_1(x) = ax + b$

Translation by $\begin{pmatrix} 1 \\ 2 \end{pmatrix}$: replace x with $(x-1)$, add 2 $\Rightarrow f_2(x) = a(x-1) + b + 2$

Reflection in $y = 0$: multiply though by $-1 \Rightarrow f_3(x) = a(-x+1) - b - 2$

Horizontal stretch, scale factor $\frac{1}{3}$: replace x with $3x \Rightarrow$

$f_4(x) = a(-3x+1) - b - 2$
$ = -3ax + a - b - 2$

$\therefore g(x) = 4 - 15x = a - b - 2 - 3ax$

Comparing coefficients of x^1:
$-3a = -15 \Rightarrow a = 5$

Comparing coefficients of x^0:
$a - b - 2 = 4 \Rightarrow b = -1$

b $f(x) = ax^2 + bx + c$

Reflection in $x = 0$: replace x with $-x$
$\Rightarrow f_2(x) = ax^2 - bx + c$

Translation by $\begin{pmatrix} -1 \\ 3 \end{pmatrix}$: replace x with $(x+1)$, add 3

$\Rightarrow f_3(x) = a(x+1)^2 - b(x+1) + c + 3$

Horizontal stretch, scale factor 2: replace x with $\frac{x}{2} \Rightarrow$

$f_4(x) = a\left(\frac{x}{2}+1\right)^2 - b\left(\frac{x}{2}+1\right) + c + 3$

$ = \frac{a}{4}x^2 + ax + a - \frac{b}{2}x - b + c + 3$

$ = \frac{a}{4}x^2 + \left(a - \frac{b}{2}\right)x + a - b + c + 3$

$\therefore g(x) = 4x^2 + ax - 6$

$ = \frac{a}{4}x^2 + \left(a - \frac{b}{2}\right)x + a - b + c + 3$

Comparing coefficients of x^2:
$\frac{a}{4} = 4 \Rightarrow a = 16$

Comparing coefficients of x^1:
$a - \frac{b}{2} = a \Rightarrow b = 0$

Comparing coefficients of x^0:
$a - b + c + 3 = -6 \Rightarrow c = -25$

Mixed examination practice 5

Short questions

1 a Graph of $y = 3f(x-2)$ is obtained from the graph of $y = f(x)$ by:

translation $\begin{pmatrix} 2 \\ 0 \end{pmatrix}$ and vertical stretch with scale factor 3

$\Rightarrow$ asymptote becomes $x = 0$;
x-intercepts become $(2, 0)$ and $(6, 0)$

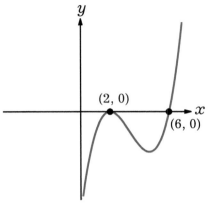

Figure 5MS.1.1

b Graph of $y = f\left(\dfrac{x}{3}\right) - 2$ is obtained from the graph of $y = f(x)$ by:
horizontal stretch with scale factor 3 and translation $\begin{pmatrix} 0 \\ -2 \end{pmatrix}$

$\Rightarrow$ asymptote becomes $x = -6$; single x-intercept at $(15, 0)$

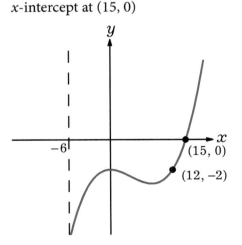

Figure 5MS.1.2

2 $f(x) = x^3 - 1$

Translation by $\begin{pmatrix} 2 \\ 0 \end{pmatrix}$: replace x with $(x-2)$

$\Rightarrow f_2(x) = (x-2)^3 - 1$

Vertical stretch with scale factor 2:
multiply by 2 $\Rightarrow f_3(x) = 2\left[(x-2)^3 - 1\right]$

So new graph is

$$y = 2\left[(x-2)^3 - 1\right]$$
$$= 2\left[x^3 - 6x^2 + 12x - 8 - 1\right]$$
$$= 2x^3 - 12x^2 + 24x - 18$$

3 a Graph of $y = f(x-1) + 2$ is obtained from the graph of $y = f(x)$ by a translation $\begin{pmatrix} 1 \\ 2 \end{pmatrix}$

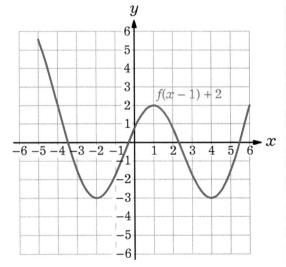

Figure 5MS.3

b Original minimum points were $(-3, -5)$ and $(3, -5)$. Adding 1 to the x-coordinates and 2 to the y-coordinates gives new minimum points of $(-2, -3)$ and $(4, -3)$.

4 Let $f(x)=(x-1)^2$, $g(x)=3(x+2)^2$

Then $g(x)=3f(x+3)$, which represents a translation by $\begin{pmatrix} -3 \\ 0 \end{pmatrix}$ and a vertical stretch with scale factor 3.

Note: there are alternative valid answers. For instance,

$g(x)=(\sqrt{3}x+2\sqrt{3})^2 = f(\sqrt{3}x+1+2\sqrt{3})$,

representing a translation $\begin{pmatrix} -1-2\sqrt{3} \\ 0 \end{pmatrix}$

followed by a horizontal stretch with scale factor $\dfrac{1}{\sqrt{3}}$.

Also, $g(x)=(\sqrt{3}x+2\sqrt{3})^2$

$= f(\sqrt{3}x+2\sqrt{3}+1)$

$= f\left(\sqrt{3}\left(x+2+\dfrac{1}{\sqrt{3}}\right)\right)$,

which represents a horizontal stretch with scale factor $\dfrac{1}{\sqrt{3}}$ followed by a translation

$\begin{pmatrix} -2-\dfrac{1}{\sqrt{3}} \\ 0 \end{pmatrix}$.

COMMENT

In questions on transformations, it is often the case with simple curves that several possible transformations will lead to the same effective change, as here. In such cases, any single valid answer is acceptable, but one is usually simpler than the others.

5 a $y=3f\left(\dfrac{x}{2}\right)$

In $y=f(x)$, x is replaced by $\dfrac{x}{2}$, corresponding to a horizontal stretch with scale factor 2.

Multiplication by 3 corresponds to a vertical stretch with scale factor 3.

b Graph of $y=\ln x$ has a vertical asymptote $x=0$ and an axis intercept $(1, 0)$.

New graph still has asymptote at $x=0$, but the intercept shifts to $(2, 0)$.

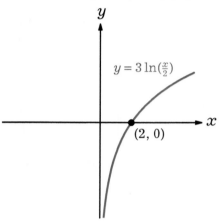

Figure 5MS.5.1

c The transformation from $y=3\ln\left(\dfrac{x}{2}\right)$ to $y=3\ln\left(\dfrac{x}{2}+1\right)=3\ln\left(\dfrac{x+2}{2}\right)$ is a translation by $\begin{pmatrix} -2 \\ 0 \end{pmatrix}$.

New graph shows the answer in (b) shifted 2 units to the left; the asymptote is at $x=-2$ and the axis intercept is $(0, 0)$.

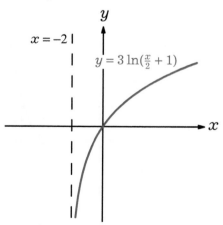

Figure 5MS.5.2

42 Mixed examination practice 5

6 The graph of $y = f(3x - 2)$ is the graph of $y = f(x)$ after a translation $\begin{pmatrix} 2 \\ 0 \end{pmatrix}$ followed by a horizontal stretch with scale factor $\frac{1}{3}$.

New graph will have an axis intercept at $(0, 0)$ but retain the asymptote $y = 2$.

It will pass through the point $\left(\frac{2}{3}, 2\right)$ [transformed from the point $(0, 2)$].

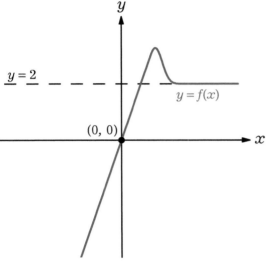

Figure 5MS.6

Long questions

1 a $y = 3x^2 - 12x + 12 = 3(x - 2)^2$

Transformation from $y = f(x)$ to $y = 3f(x - 2)$: translation by $\begin{pmatrix} 2 \\ 0 \end{pmatrix}$ and vertical stretch with scale factor 3.

b $y = x^2 + 6x - 1 = (x + 3)^2 - 10$

Transformation from $y = f(x + 3) - 10$ to $y = f(x)$: translation by $\begin{pmatrix} 3 \\ 10 \end{pmatrix}$, which is equivalent to translation by $\begin{pmatrix} 3 \\ 0 \end{pmatrix}$ and translation by $\begin{pmatrix} 0 \\ 10 \end{pmatrix}$.

> **COMMENT**
>
> It is usual to go from $y = f(x)$ to $y = f(x + 3) - 10$, in which case the transformation would be a translation by $\begin{pmatrix} -3 \\ -10 \end{pmatrix}$. However, this question asks for the transformation in the opposite direction, hence the translation by $\begin{pmatrix} 3 \\ 10 \end{pmatrix}$.

c Transformation of $y = x^2 + 6x - 1$ to $y = 3x^2 - 12x + 12$ can be achieved by the transformation in (b) followed by the transformation in (a):

translation by $\begin{pmatrix} 3 \\ 10 \end{pmatrix}$ and then

translation by $\begin{pmatrix} 2 \\ 0 \end{pmatrix}$ and vertical stretch with scale factor 3, which is equivalent to translation by $\begin{pmatrix} 5 \\ 10 \end{pmatrix}$ and vertical stretch with scale factor 3.

2 a Translation by $\begin{pmatrix} -2 \\ 0 \end{pmatrix}$

b i Graph of $y = \ln(x + 2)$ is the graph of $y = \ln x$ translated 2 units left: asymptote at $x = -2$; intercepts at $(-1, 0)$ and $(0, \ln 2)$

ii $y = \ln(x^4 + 4x + 4)$
$= \ln(x + 2)^2 = 2\ln(x + 2)$

New graph is the graph in (i) after a vertical stretch of scale factor 2: asymptote remains at $x = -2$; intercepts become $(-1, 0)$ and $(0, 2\ln 2)$

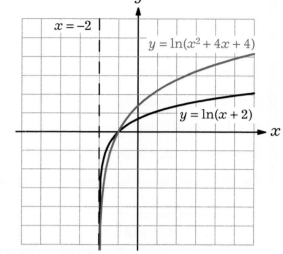

Figure 5ML.2

c i The turning point on the right has shifted from (1, 4) to (3, −4). Reflection in the x-axis has switched the sign of the y-coordinate, so the translation must be $\begin{pmatrix} 2 \\ 0 \end{pmatrix}$.

ii $g(x)$ has been translated by $\begin{pmatrix} 2 \\ 0 \end{pmatrix}$ and reflected in the x-axis, so $h(x) = -g(x-2)$.

$h(x) = -g(x-2)$
$= -\left((x-2)^3 - 2(x-2) + 5\right)$
$= -\left(x^3 - 6x^2 + 12x - 8 - 2x + 4 + 5\right)$
$= -x^3 + 6x^2 - 10x - 1$

$\therefore a = -1, \ b = 6, \ c = -10, \ d = -1$

3 a As $x \to \infty$, $f(x) \to \dfrac{3x}{x} = 3$

$\therefore$ horizontal asymptote is $y = 3$

b $f(x) = \dfrac{3x-5}{x-2}$

$= \dfrac{3(x-2)+1}{x-2}$

$= \dfrac{3(x-2)}{x-2} + \dfrac{1}{x-2}$

$= 3 + \dfrac{1}{x-2}$

$\therefore p = 3, \ q = 1$

c If $g(x) = \dfrac{1}{x}$, then $f(x) = 3 + g(x-2)$.

Transformation from g to f is a translation by $\begin{pmatrix} 2 \\ 3 \end{pmatrix}$.

d $y = \dfrac{3x-5}{x-2}$

$(x-2)y = 3x-5$
$xy - 2y = 3x - 5$
$xy - 3x = 2y - 5$
$x(y-3) = 2y - 5$
$\Rightarrow x = \dfrac{2y-5}{y-3}$

$\therefore f^{-1}(x) = \dfrac{2x-5}{x-3}$

Range of $f(x)$ is $y \neq 3$, so domain of $f^{-1}(x)$ is $x \neq 3$.

e The graph of $y = f^{-1}(x)$ is obtained from the graph of $y = f(x)$ by reflecting in the line $y = x$.

6 Sequences and series

Exercise 6A

5 a $u_1 = 1 \times 2^1 = 2$

b $\dfrac{u_{n+1}}{u_n} = \dfrac{(n+1)2^{n+1}}{n 2^n} = \dfrac{2(n+1)}{n} = 2\left(1 + \dfrac{1}{n}\right)$

6 a $u_2 = 3(2) - 2(1) = 4$
$u_3 = 3(4) - 2(2) = 8$
$u_4 = 3(8) - 2(4) = 16$

b i It appears that $u_n = 2^n$

ii If $u_n = 2^n$ then $u_{n-1} = 2^{n-1}$ and $u_{n+1} = 2^{n+1}$, so

$3u_n - 2u_{n-1} = 3(2^n) - 2(2^{n-1})$
$\qquad = 3 \times 2^n - 2^n$
$\qquad = 2^n(3-1)$
$\qquad = 2^n \times 2$
$\qquad = 2^{n+1}$
$\qquad = u_{n+1}$

i.e. $u_n = 2^n$ satisfies the equation $u_{n+1} = 3u_n - 2u_{n-1}$.

> **COMMENT**
>
> This is an example of a method called proof by induction, which in this case can be used to establish that the result $u_n = 2^n$ is true for all $n \in \mathbb{Z}^+$. Although induction is beyond the scope of Standard Level (it is on the Higher Level syllabus), a question like this which leads the student through a proof without explicit knowledge of induction could be set.

Exercise 6C

> **COMMENT**
>
> When approaching worded questions on sequences and series, it is wise to begin by setting out in equation form the information given in the question, so that you can readily refer to it and so that your notation is clear from the outset.

3 a $a_1 = 5, \ a_2 = 13$

$d = a_2 - a_1$
$\quad = 13 - 5 = 8$

$a_n = a_1 + (n-1)d$
$\quad = 5 + 8(n-1)$
$\quad = 8n - 3$

b $a_n < 400$

$8n - 3 < 400$

$n < \dfrac{403}{8} = 50.375$

So the first 50 terms are less than 400.

4 $u_{10} = 61$

$\therefore u_1 + (10-1)d = 61$

$\Rightarrow u_1 + 9d = 61 \quad \ldots (1)$

$u_{13} = 79$

$\therefore u_1 + (13-1)d = 79$

$\Rightarrow u_1 + 12d = 79 \quad \ldots (2)$

$(2) - (1):$

$3d = 18 \Rightarrow d = 6$

$u_{20} = u_{10} + 10d$
$\quad = 61 + 60 = 121$

> **COMMENT**
> Here $10d$ has been added to the tenth term to find the twentieth term, but this could also be calculated by first finding u_1 from equation (1) or (2) and then using the usual formula for u_n.

5 $u_8 = 74$

$\therefore u_1 + (8-1)d = 74$

$\Rightarrow u_1 + 7d = 74 \quad \ldots(1)$

$u_{15} = 137$

$\therefore u_1 + (15-1)d = 137$

$\Rightarrow u_1 + 14d = 137 \quad \ldots(2)$

(2) − (1):

$7d = 63 \Rightarrow d = 9$

Substituting in (1):

$u_1 + 7 \times 9 = 74 \Rightarrow u_1 = 11$

Let $u_n = 227$; then

$11 + (n-1)9 = 227$

$9n + 2 = 227$

$n = \dfrac{225}{9} = 25$

i.e. the 25th term is 227.

6 Third rung is 70 cm above ground:

$u_3 = 70$

$\therefore u_1 + (3-1)d = 70$

$\Rightarrow u_1 + 2d = 70 \quad \ldots(1)$

Tenth rung is 210 cm above ground:

$u_{10} = 210$

$\therefore u_1 + (10-1)d = 210$

$\Rightarrow u_1 + 9d = 210 \quad \ldots(2)$

(2) − (1):

$7d = 140 \Rightarrow d = 20$

Substituting in (1):

$u_1 + 2 \times 20 = 70 \Rightarrow u_1 = 30$

Top rung is 350 cm above ground, so let $u_n = 350$; then

$30 + (n-1)20 = 350$

$20n + 10 = 350$

$n = \dfrac{340}{20} = 17$

i.e. the ladder has 17 rungs.

7 $u_1 = 2$

$u_2 = a - b$

$\therefore 2 + d = a - b \quad \ldots(1)$

$u_3 = 2a + b + 7$

$\therefore 2 + 2d = 2a + b + 7 \quad \ldots(2)$

$u_4 = a - 3b$

$\therefore 2 + 3d = a - 3b \quad \ldots(3)$

(3) − (1):

$2d = a - 3b - (a - b)$

$\Rightarrow d = -\dfrac{2b}{2} = -b$

Substituting $d = -b$ in (1):

$2 - b = a - b$

$\Rightarrow a = 2$

(3) − (2):

$d = a - 3b - (2a + b + 7)$

$\therefore -b = -a - 4b - 7$

$\Rightarrow b = \dfrac{-a - 7}{3}$

$= \dfrac{-2 - 7}{3} = -3$

8 a First 9 pages are numbered with single digits for a total of 9 digits.

10th and 11th pages are each numbered with two digits.

Total number of digits for the first 11 pages is $9 + 2 \times 2 = 13$.

b First 9 pages: 9 pages at 1 digit per page $\Rightarrow$ total 9

Pages 10–99: 90 pages at 2 digits per page $\Rightarrow$ total 180

So the total number of digits on the first 99 pages is 189.

Then, since pages 100–999 have 3 digits per page, define an arithmetic sequence with first term 192 (189 plus the 3 digits on page 100) and common difference 3. Letting $u_n = 1260$:

$$192 + (n-1)3 = 1260$$
$$189 + 3n = 1260$$
$$n = 357$$

So there are $99 + 357 = 456$ pages in total.

Exercise 6D

3 $u_2 = 7$
$\Rightarrow u_1 + d = 7$ ...(1)
$S_4 = 12$
$\Rightarrow \frac{4}{2}(2u_1 + (4-1)d) = 12$
$2(2u_1 + 3d) = 12$
$\Rightarrow 2u_1 + 3d = 6$ ...(2)
$(2) - 2 \times (1)$:
$d = 6 - 14 = -8$
Substituting in (1):
$u_1 = 7 - (-8) = 15$

4 $u_1 = 2$, $d = 3$

a $S_n = \frac{n}{2}(2u_1 + (n-1)d)$
$= \frac{n}{2}(4 + 3(n-1))$
$= \frac{n}{2}(3n + 1)$

b $S_n = 1365$
$\frac{n}{2}(3n+1) = 1365$
$3n^2 + n = 2730$
$3n^2 + n - 2730 = 0$
$(3n + 91)(n - 30) = 0$
$\therefore n = 30$ (as $n \in \mathbb{Z}^+$)

5 a $S_n = 2n^2 - n$
$u_1 = S_1$
$= 2 \times 1^2 - 1$
$= 1$
$u_2 = S_2 - S_1$
$= (2 \times 2^2 - 2) - 1$
$= 5$
$u_3 = S_3 - S_2$
$= (2 \times 3^2 - 3) - 6$
$= 9$

> **COMMENT**
> Remember that if you are given a formula for S_n, then by substituting $n = 1$ you can immediately calculate $S_1 = u_1$.

b From part (a), $u_1 = 1$ and $d = 4$
$u_n = u_1 + (n-1)d$
$= 1 + (n-1) \times 4$
$= 4n - 3$

6 $u_1 = 85$, $d = -7$
$u_n < 0$
$\Rightarrow 85 + (n-1)(-7) < 0$
$-7n + 92 < 0$
$n > \frac{92}{7} = 13.14...$

So the last positive term is u_{13}.
$S_{13} = \frac{13}{2}[2 \times 85 + (13-1)(-7)] = 559$

7 $u_2 = 6$
$\Rightarrow u_1 + d = 6$ ...(1)
$S_4 = 8$
$\Rightarrow \frac{4}{2}(2u_1 + (4-1)d) = 8$
$2(2u_1 + 3d) = 8$
$\Rightarrow 2u_1 + 3d = 4$ ...(2)

(2) − 2 × (1):

$d = 4 - 12 = -8$

Substituting in (1):

$u_1 = 6 - (-8) = 14$

8 $u_1 = -6$, $d = 7$

$S_n > 1000$

$\frac{n}{2}[2(-6) + (n-1) \times 7] > 10000$

$\frac{n}{2}(7n - 19) > 10000$

$7n^2 - 19n - 20000 > 0$

$n < -52.1$ or $n > 54.8$ (roots from GDC)

So the smallest n such that $S_n > 10000$ is 55.

9 $S_n = 3n^2 - 2n$

$\frac{n}{2}(u_1 + u_n) = 3n^2 - 2n$

$\Rightarrow u_1 + u_n = 6n - 4$

$u_1 = S_1$
$= 3 - 2 = 1$

$\therefore 1 + u_n = 6n - 4$

$\Rightarrow u_n = 6n - 5$

> **COMMENT**
> Remember that if you are given a formula for S_n, then by substituting $n=1$ you can immediately calculate $S_1 = u_1$.

10 There are 12 angles in the sequence, so $n = 12$.

The angle of the largest sector is twice the angle of the smallest sector, so $u_{12} = 2u_1$.

Since the angles must add up to 360°,

$S_{12} = 360$

$\frac{12}{2}(u_1 + u_{12}) = 360$

$u_1 + 2u_1 = 60$

$3u_1 = 60$

$u_1 = 20$ ∴ smallest sector is 20°.

11 $\frac{u_5}{u_{12}} = \frac{6}{13}$

$\frac{u_1 + 4d}{u_1 + 11d} = \frac{6}{13}$

$13(u_1 + 4d) = 6(u_1 + 11d)$

$7u_1 = 14d$

$\Rightarrow u_1 = 2d$ ...(1)

$u_1 u_3 = 32$

$\Rightarrow u_1(u_1 + 2d) = 32$ ...(2)

Substituting (1) into (2):

$2d(2d + 2d) = 32$

$8d^2 = 32$

$d = \pm 2$

From (1):

$u_1 = 2(\pm 2) = \pm 4$

As all terms are positive, must have $a = 4$ and $d = 2$. So

$S_{100} = \frac{100}{2}(2a + 99d)$

$= 50(2 \times 4 + 99 \times 2)$

$= 10300$

12 We need to find

$112 + 140 + 154 + 182 + 196 + \cdots + 980 + 994$

This can be considered as the sum of two series:

$S_U = 112 + 154 + 196 + \cdots + 994$, series with $a_U = 112$, $d_U = 42$, $n_U = 22$

$S_V = 140 + 182 + 224 + \cdots + 980$, series with $a_V = 140$, $d_V = 42$, $n_V = 21$

$S_U + S_V = \frac{22}{2}[(2 \times 112) + (22 - 1) \times 42]$

$+ \frac{21}{2}[(2 \times 140) + (21 - 1) \times 42]$

$= 12166 + 11760$

$= 23926$

Alternatively, the value can be calculated as the difference of two series:

$S_W = 112 + 126 + 140 + \cdots + 994$, sum of the 64 three-digit multiples of 14

$S_X = 126 + 168 + \cdots + 966$, sum of the 21 three-digit multiples of 42 (i.e. multiples of both 21 and 14), which are excluded

$$S_W - S_X = \frac{64}{2}[(2\times 112) + (64-1)\times 14]$$
$$- \frac{21}{2}[(2\times 126) + (21-1)\times 42]$$
$$= 35\,392 - 11\,466$$
$$= 23\,926$$

$u_N < 10^{-6}$

$\Rightarrow 0.4^N < 10^{-6}$

$N \log 0.4 < -6$

$N > -\dfrac{6}{\log(0.4)}$

$\therefore N > 15.08$

The least such N is 16, so the 16th term is the first to be less than 10^{-6}.

Exercise 6E

4 $u_5 = u_2 r^3$

$\therefore u_5 = 162 \Rightarrow u_2 r^3 = 162$

i.e. $6r^3 = 162$

$r^3 = 27$

$r = 3$

$u_{10} = u_5 r^5$

$= 162 \times 3^5$

$= 39\,366$

5 $u_6 = u_3 r^3$

$\therefore u_6 = 7168 \Rightarrow u_3 r^3 = 7168$

i.e. $112 r^3 = 7168$

$r^3 = 64$

$r = 4$

$u_6 r^m = 1\,835\,008$

$7168 \times 4^m = 1\,835\,008$

$4^m = 256$

$m = 4$

$\therefore 1\,835\,008 = u_{6+4} = u_{10}$

6 Here $u_1 = \dfrac{2}{5}, r = \dfrac{2}{5}$

$\therefore u_n = u_1 r^{n-1} = \dfrac{2}{5} \times \left(\dfrac{2}{5}\right)^{n-1}$

$= \left(\dfrac{2}{5}\right)^n = 0.4^n$

7 $u_4 - u_3 = \pm \dfrac{75}{8} u_1$

$u_1(r^3 - r^2) = \pm \dfrac{75}{8} u_1$

$\Rightarrow 8r^3 - 8r^2 = \pm 75$

(discounting the trivial case $u_1 = 0$)

Using GDC to find the roots of the two possible cubics, there is one solution for each

$r = 2.5$ or -1.82.

8 $u_5 = u_3 r^2$

$\therefore u_5 = 48 \Rightarrow u_3 r^2 = 48$

i.e. $12 r^2 = 48$

$r^2 = 4$

$r = \pm 2$

$u_8 = u_5 r^3$

$= 48 \times (\pm 2)^3$

$= \pm 384$

9 $u_1 = a$

$u_3 = 9a$

$ar^2 = 9a$

$r^2 = 9$ (since $a \neq 0$)

$\therefore r = \pm 3$

$u_2 = a + 14$

$ar = a + 14$

$\pm 3a = a + 14$

$2a = 14$ or $-4a = 14$

$a = 7$ or $a = -3.5$

10

> **COMMENT**
>
> In sequences and series, the letters a, d and r usually have standard meanings, so take extra care in questions like this one which use these letters in other ways.

For the arithmetic progression:

$v_1 = a$

$v_2 = 1$

$\Rightarrow a + d = 1 \quad \ldots(1)$

$v_3 = b$

$\Rightarrow a + 2d = b \quad \ldots(2)$

$(2) - 2 \times (1)$:

$-a = b - 2$

$\Rightarrow b = 2 - a \quad \ldots(3)$

For the geometric progression:

$v_1 = 1$

$v_2 = a$

$\therefore r = a$

$v_3 = b$

$\therefore r^2 = b$

So $a^2 = b$

Hence, substituting in (3):

$a^2 = 2 - a$

$a^2 + a - 2 = 0$

$(a+2)(a-1) = 0$

$a = -2 \quad \text{or} \quad a = 1$

$a = 1 \Rightarrow b = 1$, which contradicts the requirement that $a \neq b$.

$\therefore a = -2$ and $b = 4$

11 $S_n = 4n^2 - 2n$

$u_1 = S_1$

$= 4(1^2) - 2(1)$

$= 2$

$u_2 = S_2 - S_1$

$= \left(4(2^2) - 2(2)\right) - 2$

$= 12 - 2$

$= 10$

$\therefore d = u_2 - u_1 = 8$

$u_{32} = u_1 + (32-1)d$

$= 2 + 31 \times 8$

$= 250$

Since u_2, u_m, u_{32} (i.e. 10, u_m, 250) form a geometric sequence,

$250 = 10r^2$

$r^2 = 25$

$r = \pm 5$

$\therefore u_m = 10r = 50$

(Reject $10r = -50$ since this clearly does not lie in the arithmetic sequence.)

Returning to the arithmetic sequence,

$u_m = u_1 + (m-1)d$

$= 2 + (m-1) \times 8$

$= 8m - 6$

$\therefore 50 = 8m - 6$

$m = 7$

Exercise 6F

3 a $u_n = 3 \times 5^{n+2}$

$= 3 \times 5^3 \times 5^{n-1}$

$= 375 \times 5^{n-1}$

Comparing this to the standard formula $u_n = u_1 r^{n-1}$, we find $r = 5$.

b From (a), $u_1 = 375$

$S_n = \dfrac{u_1(r^n - 1)}{r - 1}$

$= \dfrac{375}{4}(5^n - 1)$

4 $S_3 = a(1+r+r^2) = \dfrac{95}{4}$

$S_4 = a(1+r+r^2+r^3) = \dfrac{325}{8}$

Dividing gives

$\dfrac{1+r+r^2+r^3}{1+r+r^2} = \dfrac{325}{8} \times \dfrac{4}{95} = \dfrac{325}{190} = \dfrac{65}{38}$

$38(1+r+r^2+r^3) = 65(1+r+r^2)$

$38r^3 - 27r^2 - 27r - 27 = 0$

From GDC: $r = \dfrac{3}{2}$

$a(1+r+r^2) = \dfrac{95}{4}$

$a\left(1 + \dfrac{3}{2} + \left(\dfrac{3}{2}\right)^2\right) = \dfrac{95}{4}$

$\therefore a = 5$

5 $S_{15} = 29$

$\dfrac{u_1(r^{15}-1)}{r-1} = 29$

$\dfrac{6(r^{15}-1)}{r-1} = 29$

$r^{15} - 1 = \dfrac{29}{6}(r-1)$

$6r^{15} - 29r + 23 = 0$

From GDC: $r = -1.16$, 0.800 (or the false solution $r = 1$ which arises from multiplying by $r-1$).

$\therefore r = -1.16$ or 0.800 (3SF)

6 a $S_5 = S_4 + u_5$

$844 = 520 + u_1 r^4$

$\Rightarrow u_1 r^4 = 324$ ...(1)

$S_6 = S_5 + u_6$

$1330 = 844 + u_1 r^5$

$\Rightarrow u_1 r^5 = 486$ ...(2)

$(2) \div (1)$:

$\dfrac{u_1 r^5}{u_1 r^4} = \dfrac{486}{324}$

$\Rightarrow r = \dfrac{3}{2}$

b From (1):

$u_1 = \dfrac{324}{r^4} = 64$

$S_2 = u_1 \left(\dfrac{r^2-1}{r-1}\right)$

$= u_1 \left(\dfrac{(r-1)(r+1)}{r-1}\right)$

$= u_1(r+1)$

$= 64 \times \dfrac{5}{2}$

$= 160$

Exercise 6G

COMMENT

If you are told that the sum to infinity has a finite value, you can assume that $|r| < 1$ without further information.
If you are not told that the series converges, when evaluating or using S_∞ you should always explicitly show as part of your solution that $|r| < 1$.

3 $u_1 = -18$, $u_2 = 12$

$\Rightarrow r = \dfrac{12}{-18} = -\dfrac{2}{3}$

$S_\infty = \dfrac{u_1}{1-r}$

$= \dfrac{-18}{\dfrac{5}{3}}$

$= -\dfrac{54}{5} = -10.8$

4 a $u_1 = 18$

$u_1 = 18, u_4 = -\dfrac{2}{3}$

$\therefore 18r^3 = -\dfrac{2}{3}$

$r^3 = \dfrac{-\dfrac{2}{3}}{18} = -\dfrac{1}{27}$

$r = -\dfrac{1}{3}$

$S_n = \dfrac{u_1(1-r^n)}{1-r}$

$= \dfrac{18\left(1-\left(-\dfrac{1}{3}\right)^n\right)}{1-\left(-\dfrac{1}{3}\right)}$

$= 18\left(1-\left(-\dfrac{1}{3}\right)^n\right) \times \dfrac{3}{4}$

$= \dfrac{27}{2}\left(1-\left(-\dfrac{1}{3}\right)^n\right)$

b $S_\infty = \dfrac{27}{2}$

> **COMMENT**
> In part (b) the result $S_n = S_\infty(1-r^n)$ was used, which enabled the answer to be read off immediately from part (a).

5 $f(x) = 1 + 2x + (2x)^2 + (2x)^3 + \ldots$
is a geometric series with $a = 1$ and $r = 2x$

a $x = \dfrac{1}{3} \Rightarrow r = \dfrac{2}{3}$, with $|r| < 1$

$S_\infty = \dfrac{u_1}{1-r}$

$= \dfrac{1}{\left(\dfrac{1}{3}\right)} = 3$

b $x = \dfrac{2}{3} \Rightarrow r = \dfrac{4}{3}$, with $|r| > 1$

$S_\infty = \infty$ since every term of the series is positive and it does not converge.

6 a $S_2 = 15$

$\Rightarrow u_1(1+r) = 15$ …(1)

$S_\infty = 27$

$\Rightarrow \dfrac{u_1}{1-r} = 27$ …(2)

(1) ÷ (2):

$\dfrac{u_1(1+r)}{\dfrac{u_1}{1-r}} = \dfrac{15}{27}$

$(1+r)(1-r) = \dfrac{15}{27}$

$1 - r^2 = \dfrac{15}{27}$

$r^2 = \dfrac{4}{9}$

$r = \pm\dfrac{2}{3}$

Each term of the series is positive, so $r = \dfrac{2}{3}$

b From (2):

$\dfrac{u_1}{1-\dfrac{2}{3}} = 27$

$u_1 = 27 \times \dfrac{1}{3} = 9$

7 $S_\infty = 32$

$\Rightarrow \dfrac{u_1}{1-r} = 32$ …(1)

$S_4 = 30$

$\Rightarrow \dfrac{u_1}{1-r}(1-r^4) = 30$ …(2)

Substituting (1) into (2):

$32(1-r^4) = 30$

$1 - r^4 = \dfrac{15}{16}$

$r^4 = \dfrac{1}{16}$

$\therefore r = \dfrac{1}{2}$ ($r > 0$ as all terms are positive)

$$S_\infty - S_8 = 32 - 32\left(1-\left(\frac{1}{2}\right)^8\right)$$

$$= 32\left[1-\left(1-\left(\frac{1}{2}\right)^8\right)\right]$$

$$= 2^5 \times \frac{1}{2^8}$$

$$= \frac{1}{2^3} = \frac{1}{8}$$

8 $S_\infty = 1 + \left(\frac{2x}{3}\right) + \left(\frac{2x}{3}\right)^2 + \ldots$ has

$u_1 = 1, \ r = \frac{2x}{3}$

a Convergence occurs when $-1 < r < 1$:

$$-1 < \frac{2x}{3} < 1$$

$$-\frac{3}{2} < x < \frac{3}{2}$$

b $x = 1.2 \Rightarrow r = \frac{2 \times 1.2}{3} = 0.8$

$$S_\infty = \frac{u_1}{1-r}$$

$$= \frac{1}{0.2} = 5$$

9 $S_\infty = 13.5$

$\Rightarrow \dfrac{u_1}{1-r} = 13.5 \quad \ldots(1)$

$S_3 = S_\infty(1-r^3) = 13$

$\Rightarrow \dfrac{u_1}{1-r}(1-r^3) = 13 \quad \ldots(2)$

Substituting (1) into (2):

$13.5(1-r^3) = 13$

$$1 - r^3 = \frac{13}{13.5}$$

$$r^3 = \frac{1}{27}$$

$$r = \frac{1}{3}$$

Substituting into (1):

$\dfrac{u_1}{1-\dfrac{1}{3}} = 13.5$

$\Rightarrow u_1 = 13.5 \times \dfrac{2}{3} = 9$

10 This series has $u_1 = 2(4-3x), \ r = 4-3x$

a Convergence occurs when $-1 < r < 1$:

$-1 < 4 - 3x < 1$

$1 < x < \dfrac{5}{3}$

b $x = 1.2$

$\Rightarrow r = 4 - 3.6 = 0.4$

and $u_1 = 2 \times (4 - 3.6) = 0.8$

$S_n > 1.328$

$\dfrac{u_1}{1-r}(1-r^n) > 1.328$

$\dfrac{0.8}{0.6}(1-0.4^n) > 1.328$

$1 - 0.4^n > 0.996$

$0.4^n < 0.004$

$n \log 0.4 < \log 0.004$

$n > 6.03$

Require at least 7 terms for a sum greater than 1.328.

11 $r = 2^x$

a Convergence occurs when $-1 < r < 1$:

$-1 < 2^x < 1$

$x < 0$

b $S_\infty = 40$

$\dfrac{35}{1-2^x} = 40$

$1 - 2^x = \dfrac{35}{40}$

$2^x = \dfrac{1}{8}$

$x = -3$

6 Sequences and series

Exercise 6H

COMMENT

Be careful to define the terms you use; in finance questions it will often be critical whether you consider u_n to be the value at the start of year n or at the end of year n. If you are defining your own variables, always state the definitions clearly at the start of the answer.

1 Let u_n represent the balance at the start of year n.

u_n follows a geometric sequence with $u_1 = 1000$, $r = 1.03$

a 6th year interest $= u_7 - u_6$
$$= 1000 \times (1.03^6 - 1.03^5)$$
$$= 1000 \times 1.03^5 \times 0.03$$
$$= 34.78$$

The interest for the sixth year is £34.78.

b The balance after six years is the balance at the start of the seventh year, u_7

$$u_7 = 1000 \times 1.03^6 = 1194.05$$

Balance after six years is £1194.05.

2 Let u_n be Lars's salary in the nth year.

u_n follows an arithmetic sequence with $u_1 = 32\,000$, $d = 1500$

a $u_{20} = u_1 + 19d$
$$= 32\,000 + 19 \times 1500$$
$$= 60\,500$$

In the twentieth year his salary will be $60 500

b $S_n \geq 1\,000\,000$

$$\frac{n}{2}(2u_1 + (n-1)d) \geq 1\,000\,000$$

$$n(62\,500 + 1500n) \geq 2\,000\,000$$

$$15n^2 + 625n - 20\,000 \geq 0$$

$$3n^2 + 125n - 4000 \geq 0$$

Roots of this positive quadratic are 21.2 and −20.5 (from GDC)

$\therefore n > 21.2$

He will have earned more than $1 million after 22 years.

3 Let u_n be the balance at the start of year n.

u_n follows a geometric sequence with $u_1 = 5000$, $r = 1.063$

a After n full years the balance is the same as at the start of year $n+1$:
$$u_{n+1} = 5000 \times 1.063^n$$

b Balance at the end of 5 years:
$$u_6 = 5000 \times 1.063^5 = 6786.35$$

c i $5000 \times 1.063^n > 10\,000$

ii $5000 \times 1.063^n > 10\,000$

$$1.063^n > 2$$

$$n \log 1.063 > \log 2$$

$$n > \frac{\log 2}{\log 1.063} = 11.3$$

Balance will exceed $10 000 after 12 full years.

4 Let u_n be the number of seats in row n.

u_n follows an arithmetic sequence with $u_1 = 50$, $d = 200$

a $S_n \geq 8000$

$$\frac{n}{2}(2u_1 + (n-1)d) \geq 8000$$

$$n(200n - 100) \geq 16\,000$$

$$2n^2 - n - 160 \geq 0$$

Roots of this positive quadratic are 9.2 and −8.9 (from GDC)

$\therefore n > 9.2$

So 10 rows are required for there to be at least 8000 seats.

b $S_{10} = \dfrac{10}{2}(2\times 50 + 9\times 200) = 9500$

   $S_5 = \dfrac{5}{2}(2\times 50 + 4\times 200) = 2250$

   The percentage of seats in the front half (first 5 rows) is $\dfrac{2250}{9500} = 23.7\%$

5 a Balance at start of year n is $100\times 1.05^{n-1}$
   $\therefore V = 100\times 1.05^{20} = \$265.33$

   b Balance at the end of month m is
   $100\times\left(1+\dfrac{0.05}{12}\right)^m$

   $100\times\left(1+\dfrac{0.05}{12}\right)^m > 265.33$

   $\left(1+\dfrac{0.05}{12}\right)^m > 2.6533$

   $m\log\left(1+\dfrac{0.05}{12}\right) > \log 2.6533$

   $m > \dfrac{\log 2.6533}{\log\left(1+\dfrac{0.05}{12}\right)} = 234.6$

   It takes 235 months, equivalent to 19 years and 7 months.

6 Let u_n be the number of miles run on day n.
   u_n follows an arithmetic sequence with $u_1 = 1$, $d = \dfrac{1}{4}$

   a $S_n \geq 26$

   $\dfrac{n}{2}(2u_1 + (n-1)d) \geq 26$

   $n\left(\dfrac{7}{4} + \dfrac{n}{4}\right) \geq 52$

   $n^2 + 7n - 208 \geq 0$

   Roots of this positive quadratic are 11.3 and -7.8 (from GDC)

   $\therefore n > 11.3$

   After 12 days the total distance exceeds 26 miles.

   b $u_n > 26$

   $u_1 + (n-1)d > 26$

   $\dfrac{3}{4} + \dfrac{n}{4} > 26$

   $n > 4\times 26 - 3$

   $n > 101$

   On the 102nd day he runs more than 26 miles.

7 Let h_n be the height that the ball rises on the nth bounce, i.e. after hitting the ground n times.

   h_n follows a geometric sequence with $h_1 = 2\times 0.8 = 1.6$, $r = 0.8$

   a $h_4 = 1.6\times 0.8^3 = 0.8192$ metres

   b Total distance travelled at the end of bounce n is

   $t_n = 2 + 2\sum_{k=1}^{n} h_k$

   $= 2 + 2\dfrac{1.6(1-0.8^n)}{1-0.8}$

   $= 2 + 16(1-0.8^n)$

   The ball hits the ground for the 9th time at the end of bounce 8.

   $t_8 = 2 + 16(1-0.8^8) = 15.3$ metres

   COMMENT
   Note that the sum $\sum_{k=1}^{n} h_k$ is doubled because the ball goes up and down the same distance before it hits the ground again.

   c The model predicts a height of $h_{20} = 0.02$ metres; 20 cm is sufficiently small that it is likely to be absorbed by model inaccuracies and physical interactions not considered in this simple model.

8 Let u_n be the account balance at the beginning of year n, where $n = 1$ is 2010.

a $u_1 = 1000$

At the beginning of 2011,
$u_2 = 1000 \times 1.04 + 1000$

At the beginning of 2012,
$u_3 = (1000 \times 1.04 + 1000) \times 1.04 + 1000$
$= 1000 + 1000 \times 1.04 + 1000 \times 1.04^2$

b The pattern in (a) shows that u_n is the sum of a geometric sequence with $u_1 = 1000$, $r = 1.04$

Hence
$u_n = \dfrac{1000(1.04^n - 1)}{1.04 - 1}$
$= 25000(1.04^n - 1)$

c $u_n \geq 50000$

$25000(1.04^n - 1) \geq 50000$

$1.04^n - 1 \geq 2$

$1.04^n \geq 3$

$n \log 1.04 \geq \log 3$

$n \geq \dfrac{\log 3}{\log 1.04} = 28.01$

In the 29th year of saving, Samantha will have accumulated at least $50 000.

Mixed examination practice 6
Short questions

1 $u_4 = 9.6$
$\Rightarrow u_1 + 3d = 9.6$ ...(1)
$u_9 = 15.6$
$\Rightarrow u_1 + 8d = 15.6$ ...(2)
(2) − (1):
$5d = 15.6 - 9.6$
$d = \dfrac{6}{5} = 1.2$

Substituting in (1):
$u_1 = 9.6 - 3 \times 1.2 = 6$

$S_9 = \dfrac{9}{2}(u_1 + u_9)$
$= \dfrac{9}{2}(6 + 15.6)$
$= 97.2$

2 Geometric sequence with $u_1 = \dfrac{1}{3}$, $r = \dfrac{1}{3}$

$u_n = \dfrac{1}{3} \times \left(\dfrac{1}{3}\right)^{n-1} = \dfrac{1}{3^n}$

$u_n < 10^{-6}$

$\dfrac{1}{3^n} < 10^{-6}$

$3^n > 10^6$

$n \log 3 > \log 10^6$

$n > \dfrac{6}{\log 3} = 12.6$

The least such n is 13.

3 $u_5 = u_1 + 4d$ and $u_2 = u_1 + d$

$u_5 = 3u_2$

$u_1 + 4d = 3(u_1 + d)$

$2u_1 = d$

$\therefore \dfrac{d}{u_1} = 2$

4 This is the sum of two infinite geometric series:

$\displaystyle\sum_{n=0}^{\infty} \dfrac{2^n + 4^n}{6^n} = \sum_{n=0}^{\infty} \dfrac{2^n}{6^n} + \sum_{n=0}^{\infty} \dfrac{4^n}{6^n}$

$= \displaystyle\sum_{n=0}^{\infty} \left(\dfrac{2}{6}\right)^n + \sum_{n=0}^{\infty} \left(\dfrac{4}{6}\right)^n$

$= \displaystyle\sum_{n=0}^{\infty} \left(\dfrac{1}{3}\right)^n + \sum_{n=0}^{\infty} \left(\dfrac{2}{3}\right)^n$

$= \dfrac{1}{1 - \dfrac{1}{3}} + \dfrac{1}{1 - \dfrac{2}{3}}$

$= \dfrac{3}{2} + 3$

$= 4.5$

5 This is an arithmetic series with $u_1 = 301$ and $d = 7$.

To find the number of terms:
$u_n \leq 600$
$301 + (n-1) \times 7 \leq 600$
$7n - 7 \leq 299$
$n \leq \dfrac{306}{7} = 43.7$
$\therefore n = 43$

$S_{43} = \dfrac{43}{2}(2 \times 301 + (43-1) \times 7) = 19264$

6 Arithmetic sequence $\{u_n\}$ has $u_1 = 1$.
Geometric sequence $\{v_n\}$ has $v_1 = 1$.

$u_3 = v_2$
$\therefore 1 + 2d = r$ ...(1)

$u_4 = v_3$
$\therefore 1 + 3d = r^2$ ...(2)

Substituting (1) into (2):
$1 + 3d = (1 + 2d)^2$
$1 + 3d = 1 + 4d + 4d^2$
$4d^2 + d = 0$
$d(4d + 1) = 0$

So $d = 0$ (corresponding to $r = 1$ and both $\{u_n\}$ and $\{v_n\}$ being the constant sequence 1, 1, 1,...)

or $d = -\dfrac{1}{4}$ (corresponding to $r = \dfrac{1}{2}$).

7 $u_1 = \ln\left(\dfrac{a^3}{b^{\frac{1}{2}}}\right) = 3\ln a - \dfrac{1}{2}\ln b$

$u_2 = \ln\left(\dfrac{a^3}{b}\right) = 3\ln a - \ln b$

$u_3 = \ln\left(\dfrac{a^3}{b^{\frac{3}{2}}}\right) = 3\ln a - \dfrac{3}{2}\ln b$

from which it can be seen that the sequence is arithmetic, with $u_1 = 3\ln a - \dfrac{1}{2}\ln b$ and $d = -\dfrac{1}{2}\ln b$.

$\therefore S_{23} = \dfrac{23}{2}(2u_1 + 22d)$

$= \dfrac{23}{2}(6\ln a - \ln b - 11\ln b)$

$= 69\ln a - 138\ln b$

$= \ln\left(\dfrac{a^{69}}{b^{138}}\right)$

Long questions

1 a Let A_n be the amount in plan A after n years; then $\{A_n\}$ is an arithmetic sequence with $u_1 = 10800$, $d = 800$:
$A_n = 10000 + 800n$

b Let B_n be the amount in plan B after n years; then $\{B_n\}$ is a geometric sequence with $u_1 = 10500$, $r = 1.05$:
$B_n = 10000 \times 1.05^n$

c From GDC, intersection of the two graphs occurs at $n = 18.8$, so for the first 19 years $A_n > B_n$, i.e. plan A is better than plan B.

2 Let u_n be the number of bricks in row n, where row 1 is the top row.

Then $u_1 = 1$ and $u_{n+1} = u_n + 2$: this is an arithmetic sequence with $u_1 = 1$, $d = 2$.

a $u_n = 1 + (n-1) \times 2$
$= 2n - 1$

b $S_n = 36$
$\dfrac{n}{2}(2u_1 + (n-1)d) = 36$
$\dfrac{n}{2}(2 + 2n - 2) = 36$
$n^2 = 36$
$\therefore n = 6$

c $S_n = 4u_n + 4$

$\dfrac{n}{2}(2+(n-1)\times 2) = 4(2n-1)+4$

$n^2 = 8n$

$n^2 - 8n = 0$

$n(n-8) = 0$

$\therefore n = 8$ (reject $n=0$)

Hence $S_n = n^2 = 64$

3 a There are n integers on the nth line

b

TABLE 6ML.3

Line	Final integer	Equals
1	1	
2	3	$= 1 + 2$
3	6	$= 1 + 2 + 3$
4	10	$= 1 + 2 + 3 + 4$

From the table it can be seen that the final integer on the nth line is the sum of the first n integers, i.e. S_n for an arithmetic sequence with $u_1 = 1, d = 1$:

$S_n = \dfrac{n}{2}(2+(n-1))$

$= \dfrac{n(n+1)}{2}$

$= \dfrac{n^2+n}{2}$

c The first integer on the nth line must be $n-1$ less than the final integer:

$\dfrac{n(n+1)}{2} - (n-1) = \dfrac{n^2+n-2n+2}{2}$

$= \dfrac{n^2-n+2}{2}$

d The integers on the nth line form an arithmetic sequence of n consecutive values from $\dfrac{n^2-n+2}{2}$ to $\dfrac{n^2+n}{2}$, so their sum is

$S_n = \dfrac{n}{2}\left(\dfrac{n^2-n+2}{2} + \dfrac{n^2+n}{2}\right)$

$= \dfrac{n}{2}\left(\dfrac{2n^2+2}{2}\right)$

$= \dfrac{n}{2}(n^2+1)$

e $\dfrac{n}{2}(n^2+1) = 16\,400$

From GDC, $n = 32$

4 a Consider the mortgage as held in one account (A) and the payments in a separate account (B).

The mortgage account just rises at its interest rate: $A_n = 15\,000 \times 1.06^n$.

$\{A_n\}$ is a geometric sequence with $A_1 = 15\,000 \times 1.06$ and $r = 1.06$.

At the end of three years the mortgage account stands at $A_3 = 15\,0000 \times 1.06^3$.

The payments account works as $B_{n+1} = 10\,000 + 1.06B_n$, since each year interest is added to the previous payments and then a new £10 000 payment is made.

Therefore B_n is a geometric series of n terms with $a = 10\,000$ and $r = 1.06$.

At the end of three years, the payments account stands at $10\,000(1+1.06+1.06^2)$

So, after three years, the balance is $A_3 - B_3$:

$150000 \times 1.06^6 - (10000 \times 1.06^2 + 10000 \times 1.06 + 10000)$

$= 150000 \times 1.06^6 - 10000 \times 1.06^2 - 10000 \times 1.06 - 10000$

b Continuing the pattern:

Balance after n years $= A_n - B_n$

$$= 150000 \times 1.06^n - 10000 \frac{(1.06^n - 1)}{1.06 - 1}$$

$$= 150000 \times 1.06^n - 500000 \frac{(1.06^n - 1)}{3}$$

c For the balance at the end of n years to be ≤ 0, require

$$150000 \times 1.06^n - 500000 \frac{(1.06^n - 1)}{3} \leq 0$$

$$150000 \times 1.06^n \leq 500000 \frac{(1.06^n - 1)}{3}$$

$0.9 \times 1.06^n \leq 1.06^n - 1$

$1.06^n \geq 10$

$n \geq \dfrac{1}{\log 1.06} = 39.5$

So the mortgage will be paid off after 40 years.

7 Binomial expansion

Exercise 7A

3 The indices of each term in the expansion must sum to $n=6$; therefore there is no $x^5 y^3$ term in the expansion, so the coefficient is zero.

Exercise 7B

> **COMMENT**
>
> It is wise always to lay out an answer using the general term of the expansion first, before evaluating indices and coefficients. This helps to avoid simple errors with signs and the application of indices, and also makes the working easy to understand.

4 $(3+2x)^5$ has terms of the form $\binom{5}{r}(3)^{5-r}(2x)^r$

Require coefficient of x^4, so $r = 4$

The term is $\binom{5}{4} 3^1 (2x)^4 = 5 \times 3 \times 16 x^4 = 240 x^4$

∴ coefficient is 240.

5 a $(2+x)^n = 32 + ax + \ldots$

$\binom{n}{0}(2)^n (x)^0 + \binom{n}{1}(2)^{n-1}(x)^1 + \ldots = 32 + ax + \ldots$

$2^n + n 2^{n-1} x + \ldots = 32 + ax + \ldots$

Equating coefficients of x^0:

$2^n = 32$

∴ $n = 5$

b Equating coefficients of x^1:

$n \times 2^{n-1} = a$

∴ $a = 5 \times 2^4 = 80$

6 a $(1+2x)^n = 1+20x+ax^2+\ldots$

$\binom{n}{0}(1)^n + \binom{n}{1}(1)^{n-1}(2x)^1 + \binom{n}{2}(1)^{n-2}(2x)^2 + \ldots = 1+20x+ax^2+\ldots$

$1+n(2x)+\dfrac{n(n-1)}{2}(2x)^2+\ldots = 1+20x+ax^2+\ldots$

$1+2nx+2n(n-1)x^2+\ldots = 1+20x+ax^2+\ldots$

Equating coefficients of x^1:

$2n = 20$

$\therefore n = 10$

b Equating coefficients of x^2:

$2n(n-1) = a$

$\therefore a = 2\times 10 \times 9 = 180$

Exercise 7C

3 a $(3-5x)^4 = 1\times(3)^4 + 4\times(3)^3(-5x) + 6\times(3)^2(-5x)^2 + \ldots$

$= 81 - 540x + 1350x^2 + \ldots$

b Require that $3-5x = 2.995$, so $x = 0.001$

$x^0 = 1 \quad \Rightarrow \quad 81x^0 \quad = \quad 81$

$x^1 = 0.001 \quad \Rightarrow \quad -540x^1 \quad = \quad -0.54$

$x^2 = 0.000001 \quad \Rightarrow \quad 1350x^2 \quad = \quad 0.00135$

Hence, approximately, $2.995^4 = 81 - 0.54 + 0.00135 = 80.46135$

> **COMMENT**
> In this type of question, find a value of x which makes the first part of the question relevant to finding the approximation. In more complicated questions this may require some ingenuity.

4 $(y+3y^2)^6 = y^6(1+3y)^6$

$= y^6\left[1\times 1^6 + 6\times 1^5(3y) + 15\times 1^4(3y)^2 + 20\times 1^3(3y)^3 + \ldots\right]$

$= y^6 + 18y^7 + 135y^8 + 540y^9 + \ldots$

5 General term of $(x-2y)^5$ has the form $\binom{5}{r}x^{5-r}(-2y)^r$

Coefficient of this term is $\binom{5}{r}(-2)^r$

a $\binom{5}{r}(-2)^r = 40$

$\Rightarrow r = 2$

$\therefore$ term is $\binom{5}{2}x^3(-2y)^2 = 40x^3y^2$

b $\binom{5}{r}(-2)^r = -80$

$\Rightarrow r = 3$

Term is $\binom{5}{3}x^2(-2y)^3 = -80x^2y^3$

6 General term of $(1-5x)^9$ has the form $\binom{9}{r}1^{9-r}(-5x)^r$

Require coefficient of x^3, so $r = 3$.

Term is $\binom{9}{3}(-5x)^3 = 84 \times (-125x^3) = -10500x^3$

Coefficient is $-10\,500$

7 General term of $(3-2x)^7$ has the form $\binom{7}{r}(3)^{7-r}(-2x)^r$

Require x^2 term, so $r = 2$.

Term is $\binom{7}{2}(3)^5(-2x)^2 = 21 \times 243 \times 4x^2 = 20412x^2$

8 General term of $(3x+2y^2)^5$ has the form $\binom{5}{r}(3x)^{5-r}(2y^2)^r$

Require coefficient of x^2y^6, so $r = 3$.

Term is $\binom{5}{3}(3x)^2(2y^2)^3 = 10 \times 9x^2 \times 8y^6 = 720x^2y^6$

Coefficient is 720

9 General term of $(x+x^{-1})^8$ has the form $\binom{8}{r}(x)^{8-r}(x^{-1})^r = \binom{8}{r}x^{8-2r}$

Require coefficient of x^2, so $8 - 2r = 2$ and hence $r = 3$.

Term is $\binom{8}{3}x^2 = 56x^2$

Coefficient is 56

10 a $(2+3x)^7 = \binom{7}{0}(2)^7 + \binom{7}{1}(2)^6(3x)^1 + \binom{7}{2}(2)^5(3x)^2 + \ldots$

$= 128 + 1344x + 6048x^2 + \ldots$

Topic 7C Applying the binomial theorem

b i Require that $2+3x = 2.3$, so $x = 0.1$.

$x^0 = 1$	$\Rightarrow$ $128x^0$	$=$	128
$x^1 = 0.1$	$\Rightarrow$ $1344x^1$	$=$	134.4
$x^2 = 0.01$	$\Rightarrow$ $6048x^2$	$=$	60.48

Hence, approximately, $2.3^7 = 128 + 134.4 + 60.48 = 322.88$

ii Require $2+3x = 2.03$, so $x = 0.01$

$x^0 = 1$	$\Rightarrow$ $128x^0$	$=$	128
$x^1 = 0.01$	$\Rightarrow$ $1344x^1$	$=$	13.44
$x^2 = 0.0001$	$\Rightarrow$ $6048x^2$	$=$	0.6048

Hence, approximately, $2.03^7 = 128 + 13.44 + 0.6048 = 142.0448$

b Approximation (ii) will be more accurate, on both an absolute and a relative basis, since the discarded terms (higher powers of x) reduce more rapidly in this case and are less significant to the total.

11 a General term of $(e + 2e^{-1})^5$ has the form $\binom{5}{r}(e)^{5-r}(2e^{-1})^r = \binom{5}{r}2^r e^{5-2r}$

$\therefore (e + 2e^{-1})^5 = 1 \times e^5 + 5 \times 2e^3 + 10 \times 4e^1 + 10 \times 8e^{-1} + 5 \times 16e^{-3} + 1 \times 32e^{-5}$

$$= e^5 + 10e^3 + 40e + \frac{80}{e} + \frac{80}{e^3} + \frac{32}{e^5}$$

b The expansion of $(e - 2e^{-1})^5$ is the same as the expansion of $(e + 2e^{-1})^5$ except that alternate terms will now be negative:

$$(e - 2e^{-1})^5 = e^5 - 10e^3 + 40e - \frac{80}{e} + \frac{80}{e^3} - \frac{32}{e^5}$$

$$\therefore (e + 2e^{-1})^5 + (e - 2e^{-1})^5 = 2e^5 + 80e + \frac{160}{e^3}$$

12 General term of $(x - ay)^n$ has the form $\binom{n}{r}(x)^{n-r}(ay)^r$

a Require $\binom{n}{r}(x)^{n-r}(ay)^r = 60x^4 y^2$

By inspecting the indices: $n = 4 + 2 = 6$

b By inspecting the index of y: $r = 2$

Require $\binom{6}{2}x^4(ay)^2 = 60x^4 y^2$

$\therefore 15a^2 x^4 y^2 = 60x^4 y^2$

$15a^2 = 60$

$a^2 = 4$

$a = \pm 2$

13 $(2z^2+3z^{-1})^4 = \binom{4}{0}(2z^2)^4 + \binom{4}{1}(2z^2)^3(3z^{-1})^1 + \binom{4}{2}(2z^2)^2(3z^{-1})^2$
$\qquad + \binom{4}{3}(2z^2)^1(3z^{-1})^3 + \binom{4}{4}(3z^{-1})^4$

$= 1 \times 16z^8 + 4 \times 8z^6 \times 3z^{-1} + 6 \times 4z^4 \times 9z^{-2} + 4 \times 2z^2 \times 27z^{-3} + 1 \times 81z^{-4}$

$= 16z^8 + 96z^5 + 216z^2 + 216z^{-1} + 81z^{-4}$

14 a $(1+x)^n(1-x)^n = [(1+x)(1-x)]^n$
$\qquad = (1-x^2)^n$

b $(1-x)^{10}(1+x)^{10} = (1-x^2)^{10}$

$= \binom{10}{0}(1)^{10} + \binom{10}{1}(1)^9(-x^2)^1 + \binom{10}{2}(1)^8(-x^2)^2 + \ldots$

$= 1 - 10x^2 + 45x^4 + \ldots$

15 a $(3x^2y + 5xy^{-1})^n = 27x^6y^3 + 135x^5y + \ldots$

$\binom{n}{0}(3x^2y)^n + \ldots = 27x^6y^3 + 135x^5y + \ldots$

Equating the first terms:

$\binom{n}{0}(3x^2y)^n = 27x^6y^3$

$\Rightarrow n = 3$

b $(3x^2y + 5xy^{-1})^3 = \binom{3}{0}(3x^2y)^3 + \binom{3}{1}(3x^2y)^2(5xy^{-1})^1$

$\qquad + \binom{3}{2}(3x^2y)^1(5xy^{-1})^2 + \binom{3}{3}(5xy^{-1})^3$

$= 3^3x^6y^3 + 3 \times 9 \times 5x^5y + 3 \times 3 \times 25x^4y^{-1} + 5^3x^3y^{-3}$

$= 27x^6y^3 + 135x^5y + 225x^4y^{-1} + 125x^3y^{-3}$

16 General term of $(2x + x^{-\frac{1}{2}})^5$ has the form $\binom{5}{r}(2x)^{5-r}(x^{-\frac{1}{2}})^r = \binom{5}{r}2^{5-r}x^{5-\frac{3r}{2}}$

Require $5 - \dfrac{3r}{2} = 2$, so $r = 2$

Term is $\binom{5}{2}2^3x^2 = 80x^2$

Coefficient is 80

17 General term of $(x - 2x^{-2})^9$ has the form $\binom{9}{r}(x)^{9-r}(-2x^{-2})^r = \binom{9}{r}(-2)^r x^{9-3r}$

Require $9 - 3r = 0$, so $r = 3$

Term is $\binom{9}{3}(-2)^3 x^0 = -672$

Constant coefficient is -672

18 General term of $(x^2 - 3x^{-1})^7$ has the form $\binom{7}{r}(x^2)^{7-r}(-3x^{-1})^r = \binom{7}{r}(-3)^r x^{14-3r}$

Require $14 - 3r = 5$, so $r = 3$

Term is $\binom{7}{3}(-3)^3 x^5 = -945x^5$

19 General term of $(2x - 5x^{-2})^{12}$ has the form $\binom{12}{r}(2x)^{12-r}(-5x^{-2})^r = \binom{12}{r}2^{12-r}(-5)^r x^{12-3r}$

Require $12 - 3r = 0$, so $r = 4$

Term independent of x is $\binom{12}{4}2^8(-5)^4 x^0 = 79\,200\,000$

20 General term of $(1+x)^7$ has the form $\binom{7}{r}x^r$

Term in x^5 in $(1+3x)(1+x)^7$ will be $1 \times \binom{7}{5}x^5 + 3x \times \binom{7}{4}x^4 = 126x^5$

Coefficient of x^5 is 126

21 $(1+ax)^n = 1 + 10x + 40x^2 + \ldots$ means that

$\binom{n}{0}(1)^n + \binom{n}{1}(1)^{n-1}(ax)^1 + \binom{n}{2}(1)^{n-2}(ax)^2 + \ldots$

$= 1 + 10x + 40x^2 + \ldots$

Comparing coefficients:

$\binom{n}{1}ax = 10x$

$anx = 10x$

$a = \dfrac{10}{n}$

$\binom{n}{2}a^2 x^2 = 40x^2$

$\dfrac{n(n-1)a^2}{2}x^2 = 40x^2$

Substituting $a = \dfrac{10}{n}$:

$$\frac{50(n-1)}{n} = 40$$

$$50n - 50 = 40n$$

$$10n = 50$$

$$n = 5$$

$$\therefore a = 2$$

Mixed examination practice 7

Short questions

1 General term of $(2-x)^{12}$ has the form $\binom{12}{r} 2^{12-r}(-x)^r$

Term in x^5 is $\binom{12}{5}(2)^7(-x)^5 = 792 \times 128 \times (-x^5) = -101\,376 x^5$

Coefficient is $-101\,376$

2 $(2x^{-1} + 5y)^3 = \binom{3}{0}(2x^{-1})^3 + \binom{3}{1}(2x^{-1})^2(5y) + \binom{3}{2}(2x^{-1})(5y)^2 + \binom{3}{3}(5y)^3$

$= 8x^{-3} + 3(4x^{-2})(5y) + 3(2x^{-1})(25y^2) + 125y^3$

$= 8x^{-3} + 60x^{-2}y + 150x^{-1}y^2 + 125y^3$

3 $(2-\sqrt{2})^5 = \binom{5}{0}(2)^5 + \binom{5}{1}(2)^4(-\sqrt{2})^1 + \binom{5}{2}(2)^3(-\sqrt{2})^2$

$+ \binom{5}{3}(2)^2(-\sqrt{2})^3 + \binom{5}{4}(2)^1(-\sqrt{2})^4 + \binom{5}{5}(-\sqrt{2})^5$

$= 32 + 5(16)(-\sqrt{2}) + 10(8)(2) + 10(4)(-2\sqrt{2}) + 5(2)(4) - 4\sqrt{2}$

$= 232 - 164\sqrt{2}$

4 General term is $\binom{4}{r}(x^3)^r(-2x^{-1})^{4-r} = \binom{4}{r}(-2)^{4-r} x^{4r-4}$

Term in x^0 (the constant term) occurs when $r = 1$, so the coefficient is $\binom{4}{1}(-2)^3 = -32$

5 $(x^2 - 2x^{-1})^4 = \binom{4}{0}(x^2)^4 + \binom{4}{1}(x^2)^3(-2x^{-1}) + \binom{4}{2}(x^2)^2(-2x^{-1})^2$

$+ \binom{4}{3}(x^2)^1(-2x^{-1})^3 + \binom{4}{4}(-2x^{-1})^4$

$= x^8 - 8x^5 + 24x^2 - 32x^{-1} + 16x^{-4}$

6 General term of $(c+d)^{14}$ has the form $\binom{14}{r}c^{14-r}d^r$

In $(2c+5d)(c+d)^{14}$, the term in c^4d^{11} arises from $2c\binom{14}{11}c^3d^{11}+5d\binom{14}{10}c^4d^{10}$

Coefficient is $2\times 364+5\times 1001=5733$

7 General term of $(1+x)^5$ has the form $\binom{5}{r}x^r$

In $(1-x^2)(1+x)^5$, the term in x^6 arises from $(-x^2)\binom{5}{4}x^4=-5x^6$

Coefficient is -5

Long questions

1 a The graph of $y=(x+2)^3$ is the graph of $y=x^3$ after a translation by $\binom{-2}{0}$.
Axis intercepts are at $(-2, 0)$ and $(0, 8)$.

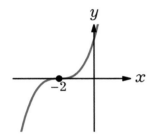

Figure 7ML.1 Graph of $y = (x+2)^3$

b $(x+2)^3 = \binom{3}{0}(x)^3 + \binom{3}{1}(x)^2(2) + \binom{3}{2}(x)(2)^2 + \binom{3}{3}(2)^3$

$= x^3+6x^2+12x+8$

c Require that $x+2=2.001$, so take $x=0.001$

$x^0 = 1 \quad\Rightarrow\quad 8x^0 = 8$

$x^1 = 0.001 \quad\Rightarrow\quad 12x^1 = 0.012$

$x^2 = 0.000001 \quad\Rightarrow\quad 6x^2 = 0.000006$

$x^3 = 0.000000001$

Hence, $2.001^3 = 8+0.012+0.000006+0.000000001 = 8.012006001$

d $x^3+6x^2+12x+16=0$

$(x+2)^3+8=0$

$(x+2)^3=-8$

$x+2=-2$

$x=-4$

2 a Total of the indices is $n = 3 + 2 = 5$

   b General term of $(2x + ay)^n$ is $\binom{n}{r}(2x)^{n-r}(ay)^r$

   The y index is $2 \Rightarrow r = 2$

   Term is $\binom{5}{2}(2x)^3(ay)^2 = 80a^2x^3y^2$

   $\therefore 80a^2x^3y^2 = 20x^3y^2$

   $a^2 = \dfrac{1}{4}$

   $a = \pm\dfrac{1}{2}$

   c With $a = \dfrac{1}{2}$,

   $\left(2x + \dfrac{1}{2}y\right)^5 = \binom{5}{0}(2x)^5 + \binom{5}{1}(2x)^4\left(\dfrac{1}{2}y\right) + \binom{5}{2}(2x)^3\left(\dfrac{1}{2}y\right)^2 + \binom{5}{3}(2x)^2\left(\dfrac{1}{2}y\right)^3 + \ldots$

   $= 32x^5 + 40x^4y + 20x^3y^2 + 5x^2y^3 + \ldots$

   d Require $2x + \dfrac{1}{2}y = 20.05$; take $x = 10$, $y = 0.1$:

   $32x^5 = 3\,200\,000$

   $40x^4y = 40\,000$

   $20x^3y^2 = 200$

   $5x^2y^3 = 0.5$

   $\therefore 20.05^5 \approx 3\,200\,000 + 40\,000 + 200 + 0.5$

   $\approx 3\,240\,200$ to the nearest hundred

8 Circular measure and trigonometric functions

Exercise 8B

13 $\cos(\pi+x)+\cos(\pi-x)$
$= \cos\pi\cos x - \sin\pi\sin x + \cos\pi\cos x + \sin\pi\sin x$
$= -\cos x - 0 - \cos x + 0$
$= -2\cos x$

14 $\sin x + \sin\left(x+\dfrac{\pi}{2}\right) + \sin(x+\pi) + \sin\left(x+\dfrac{3\pi}{2}\right) + \sin(x+2\pi)$

$= \sin x + \sin x \cos\dfrac{\pi}{2} + \cos x \sin\dfrac{\pi}{2} + \sin x \cos\pi + \cos x \sin\pi$

$\quad + \sin x \cos\dfrac{3\pi}{2} + \cos x \sin\dfrac{3\pi}{2} + \sin x$

$= \sin x + 0 + \cos x - \sin x + 0 + 0 - \cos x + \sin x$

$= \sin x$

> **COMMENT**
> Remember that $\sin(x+2\pi) = \sin x$ by the periodicity of the sine function, so it isn't necessary to expand the last term in the expression.

Exercise 8E

4 $y = p\sin(qx)$ has amplitude p, i.e. y ranges from $-p$ to p.

From the graph, $p = 5$

$y = p\sin(qx)$ has period $\dfrac{2\pi}{q}$, so the second positive zero occurs at $x = \dfrac{2\pi}{q}$

From the graph, $\dfrac{2\pi}{q} = \pi \Rightarrow q = 2$

5 $y = a\cos(x-b)$ has amplitude a, i.e. y ranges from $-a$ to a.

From the graph, $a = 2$

$y = a\cos(x-b)$ has a zero at $x = 90° + b$

From the graph, the smallest positive solution is $110°$, so $b = 20°$

6 a $y = 1 + \sin 2x$: amplitude 1; centre $y = 1$; period π; axis intercepts

$(0, 1)$, $\left(\dfrac{3\pi}{4}, 0\right)$, $\left(\dfrac{7\pi}{4}, 0\right)$

$y = 2\cos x$: amplitude 2; centre $y = 0$; period 2π; axis intercepts

$(0, 2)$, $\left(\dfrac{\pi}{2}, 0\right)$, $\left(\dfrac{3\pi}{2}, 0\right)$

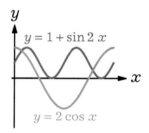

Figure 8E.6

b From Figure 8E.6, there are two points of intersection in $[0, 2\pi]$,
∴ two solutions.

c The pattern of the two curves in Figure 8E.6 repeats every 2π.
Since there are 2 solutions in an interval of length 2π, there must be 8 solutions in an interval of 8π.

7 a $y = 2\cos(x + 60°)$: amplitude 2; centre $y = 0$; period π; axis intercepts $(0, 1)$, $(30°, 0)$, $(210°, 0)$

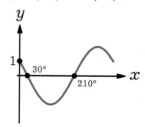

Figure 8E.7

b At maximum point:
$\cos(x + 60°) = 1$
$\Rightarrow x + 60° = 0°, 360°$
∴ $x = 300°$

At minimum point:
$\cos(x + 60°) = -1$
$\Rightarrow x + 60° = 180°$
∴ $x = 120°$

The minimum and maximum points are $(120°, -2)$ and $(300°, 2)$

c The graph of $y = 2\cos(x + 60°) - 1$ is the graph of $y = 2\cos(x + 60°)$ after a translation by $\begin{pmatrix} 0 \\ -1 \end{pmatrix}$, so its minimum and maximum points are $(120°, -3)$ and $(300°, 1)$.

Exercise 8F

1 a High tide will occur when d is a maximum, which occurs when

$\sin\left(\dfrac{\pi t}{12}\right) = 1$

$\dfrac{\pi t}{12} = \dfrac{\pi}{2}$

$t = 6$

Low tide will occur when d is a minimum, which occurs when

$\sin\left(\dfrac{\pi t}{12}\right) = -1$

$\dfrac{\pi t}{12} = \dfrac{3\pi}{2}$

$t = 18$

At high tide, $d = 16 + 7 = 23$ metres
At low tide, $d = 16 - 7 = 9$ metres

b $t = 6$ corresponds to 6 hours after midnight, i.e. 6 a.m.

2 $y = a\cos(bt) + m$: amplitude a, period $\frac{2\pi}{b}$, centre $y = m$

From the graph:

amplitude $= \frac{3}{2} \Rightarrow a = \frac{3}{2}$

period $= 12 \Rightarrow \frac{2\pi}{b} = 12 \Rightarrow b = \frac{\pi}{6}$

centre $= \frac{3+6}{2} \Rightarrow m = \frac{9}{2}$

3 a $h = a\sin(kt)$: amplitude a, period $\frac{2\pi}{k}$

From the given information:

amplitude $= 5 \Rightarrow a = 5$

period $= 10 \Rightarrow \frac{2\pi}{k} = 10 \Rightarrow k = \frac{\pi}{5}$

b 3 cm below the x-axis corresponds to $h = -3$

$5\sin\left(\frac{\pi t}{5}\right) = -3$

$\sin\left(\frac{\pi t}{5}\right) = -\frac{3}{5}$

$\frac{\pi t}{5} = \sin^{-1}\left(-\frac{3}{5}\right) = 3.79,\ 5.64\ (3\text{SF})$

$\therefore t = 6.02,\ 8.98$

The point is 3 cm below the x-axis 6.02 seconds and 8.98 seconds after starting.

4 $h = 120 - 10\cos 400t$: amplitude 10, centre $h = 120$, period $\frac{2\pi}{400} = \frac{\pi}{200}$

a Greatest height is $120 - (-10) = 130$ cm; least height is $120 - 10 = 110$ cm

b The time required to complete one full oscillation is the period,

$\frac{\pi}{200} = 0.0157$ seconds

c Greatest height occurs when $\cos 400t = -1$

$400t = \pi$

$t = \frac{\pi}{400} = 0.00785\ (3\text{SF})$

i.e. 0.00785 seconds after release.

Mixed examination practice 8
Short questions

1 $y = a\sin b(x+c) + d$ has amplitude a and period/wavelength $\frac{2\pi}{b}$

a $a = 1.4$, so amplitude is 1.4 metres

b Distance between consecutive peaks is the wavelength, which is

$\frac{2\pi}{3} = 2.09$ metres (3SF)

2 $y = a\cos(bt)$ has amplitude a and period $\frac{2\pi}{b}$

a Period is $\frac{2\pi}{0.08} = 25\pi = 78.5$ seconds (3SF)

b The amplitude is equivalent to the radius of the track, i.e. 60 metres.

$\therefore$ Track length $= 2\pi \times 60$

$= 120\pi$

$= 377$ metres (3SF)

c Speed $= \dfrac{\text{Distance}}{\text{Time}}$

$= \dfrac{120\pi}{25\pi}$

$= 4.8\ \text{m s}^{-1}$

3 $f(x) = a\sin b(x+c)$ has amplitude a and period $\frac{2\pi}{b}$

a $b = 2$, so period is $\frac{2\pi}{2} = \pi$

b $x \in [0, 2\pi] \Rightarrow 2\left(x - \frac{\pi}{3}\right) \in \left[-\frac{2\pi}{3}, \frac{10\pi}{3}\right]$

$f(x) = 0$

$3\sin 2\left(x - \frac{\pi}{3}\right) = 0$

$\Rightarrow 2\left(x - \frac{\pi}{3}\right) = 0,\ \pi,\ 2\pi,\ 3\pi$

$\Rightarrow x = \frac{\pi}{3},\ \frac{5\pi}{6},\ \frac{4\pi}{3},\ \frac{11\pi}{6}$

$\therefore$ zeros are

$\left(\frac{\pi}{3}, 0\right), \left(\frac{5\pi}{6}, 0\right), \left(\frac{4\pi}{3}, 0\right), \left(\frac{11\pi}{6}, 0\right)$

c Graph of $y = 3\sin 2\left(x - \dfrac{\pi}{3}\right)$ is

maximum when $2\left(x - \dfrac{\pi}{3}\right) = \dfrac{\pi}{2} + 2k\pi$,

i.e. $x = \dfrac{7\pi}{12}, \dfrac{19\pi}{12}$; minimum when

$2\left(x - \dfrac{\pi}{3}\right) = -\dfrac{\pi}{2} + 2k\pi$, i.e. $x = \dfrac{\pi}{12}, \dfrac{13\pi}{12}$;

amplitude 3; y-intercept $\left(0, -\dfrac{3\sqrt{3}}{2}\right)$

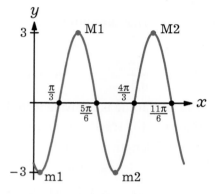

M1: $\left(\dfrac{7\pi}{12}, 3\right)$ M2: $\left(\dfrac{19\pi}{12}, 3\right)$
m1: $\left(\dfrac{\pi}{12}, -3\right)$ m2: $\left(\dfrac{13\pi}{12}, -3\right)$

Figure 8MS.3 Graph of $y = 3\sin 2\left(x - \dfrac{\pi}{3}\right)$

4 $y = a\sin(bx)$ has maximum point at $\left(\dfrac{\pi}{2b}, a\right)$

From the graph, the maximum is at $(2, 5)$

$\therefore \dfrac{\pi}{2b} = 2 \Rightarrow b = \dfrac{\pi}{4}$

and $a = 5$.

Long questions

1 a i $y = \sin(x - k) + c$ has a maximum at $\left(\dfrac{\pi}{2} + k, c + 1\right)$

By symmetry, point A is midway horizontally between the first two zeros, i.e. at $x = \dfrac{2\pi}{3}$,

$\therefore \dfrac{\pi}{2} + k = \dfrac{2\pi}{3}$

$\Rightarrow k = \dfrac{\pi}{6}$

The graph goes through the origin,

$\therefore \sin\left(0 - \dfrac{\pi}{6}\right) + c = 0$

$-\dfrac{1}{2} + c = 0$

$\Rightarrow c = \dfrac{1}{2}$

So the coordinates of A are $\left(\dfrac{2\pi}{3}, \dfrac{3}{2}\right)$

ii From (i), $k = \dfrac{\pi}{6}, c = \dfrac{1}{2}$

b Period of $y = \sin(x - k) + c$ is 2π, so the zeros are those shown in the question and the same at intervals of 2π. Within $[-4\pi, 0]$, these are

$-4\pi, -\dfrac{8\pi}{3}, -2\pi, -\dfrac{2\pi}{3}, 0$

c i For the equation $\sin\left(x - \dfrac{\pi}{6}\right) + \dfrac{1}{2} = k$, as $k < 0$, the first pair of solutions will be in the interval $\left[\dfrac{4\pi}{3}, 2\pi\right]$, and subsequent solutions will be at multiples of 2π further on, i.e. in the intervals $\left[\dfrac{10\pi}{3}, 4\pi\right]$, $\left[\dfrac{16\pi}{3}, 6\pi\right]$ and $\left[\dfrac{22\pi}{3}, 8\pi\right]$. So there are only 8 solutions in $[0, 9\pi]$.

COMMENT

It is important to check that the next such interval is not needed too: in this case $\left[\dfrac{28\pi}{3}, 10\pi\right]$ is wholly outside $[0, 9\pi]$, so all the relevant intervals have been found.

ii Given that the smallest positive solution is α, the next solution, by symmetry about $x = \dfrac{5\pi}{3}$, must be

$$2\pi - \left(\alpha - \dfrac{4\pi}{3}\right) = \dfrac{10\pi}{3} - \alpha.$$

The following solution, by periodicity, must be $2\pi + \alpha$.

So the next two solutions after α are $\dfrac{10\pi}{3} - \alpha$ and $2\pi + \alpha$.

2 i and **ii**
Using GDC as necessary:

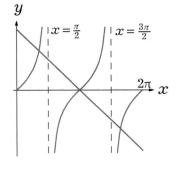

Figure 8ML.2 Graphs of $y = \tan x$ and $y = \pi - x$

b i $x + \tan x = \pi \Rightarrow \tan x = \pi - x$, so solutions of $x + \tan x = \pi$ are intersections of the graphs in Figure 8ML.2.

Given that x_0 is the first positive solution, by symmetry the other solutions in $[0, 2\pi]$ must be π and $2\pi - x_0$.

ii Since each period of $y = \tan x$ extends infinitely in the positive and negative y directions, the line $y = \pi - x$ must intersect each period of $y = \tan x$, so there are infinitely many solutions.

c i
$$\cos\left(\dfrac{\pi}{2} - A\right) = \cos\left(\dfrac{\pi}{2}\right)\cos A + \sin\left(\dfrac{\pi}{2}\right)\sin A$$
$$= 0 + \sin A$$
$$= s$$

$$\sin\left(\dfrac{\pi}{2} - A\right) = \sin\left(\dfrac{\pi}{2}\right)\cos A - \cos\left(\dfrac{\pi}{2}\right)\sin A$$
$$= \cos A - 0$$
$$= c$$

ii $\tan\left(\dfrac{\pi}{2} - A\right) = \dfrac{\sin\left(\dfrac{\pi}{2} - A\right)}{\cos\left(\dfrac{\pi}{2} - A\right)}$

$$= \dfrac{c}{s} \text{ from (i)}$$

$$= \dfrac{1}{s/c}$$

$$= \dfrac{1}{\tan A}$$

iii Let $\tan A = t$:

$$\tan A + \tan\left(\dfrac{\pi}{2} - A\right) = \dfrac{4}{\sqrt{3}}$$

$$t + \dfrac{1}{t} = \dfrac{4}{\sqrt{3}}$$

$$t^2 - \dfrac{4}{\sqrt{3}}t + 1 = 0$$

$$\left(t - \sqrt{3}\right)\left(t - \dfrac{1}{\sqrt{3}}\right) = 0$$

$\therefore \tan A = \sqrt{3}$ or $\dfrac{1}{\sqrt{3}}$

iv If $A \in \left]0, \dfrac{\pi}{2}\right[$,

$\tan A = \sqrt{3} \Rightarrow A = \dfrac{\pi}{3}$

$\tan A = \dfrac{1}{\sqrt{3}} \Rightarrow A = \dfrac{\pi}{6}$

$\therefore$ the values are $A = \dfrac{\pi}{3}$ or $\dfrac{\pi}{6}$

3 **a** Minimum value of $\cos x$ is -1, and the smallest positive value of x for which this occurs is $x = \pi$.

b i $f(x)$ to $2f\left(x+\dfrac{\pi}{6}\right)$: translation by $\begin{pmatrix} -\dfrac{\pi}{6} \\ 0 \end{pmatrix}$ and vertical stretch with scale factor 2.

ii Applying the two transformations in (i) to the minimum of $\cos x$:

the minimum point of $y = 2\cos\left(x+\dfrac{\pi}{6}\right)$ is $\left(\pi-\dfrac{\pi}{6}, -1\times 2\right) = \left(\dfrac{5\pi}{6}, -2\right)$,

i.e. the minimum value is -2, and it occurs at $x = \dfrac{5\pi}{6}$.

c i Vertical asymptotes occur where the denominator is zero.

The minimum value of the denominator $2\cos\left(x+\dfrac{\pi}{6}\right)+3$ is $-2+3=1$, so there are no vertical asymptotes.

ii The maximum denominator value is $2+3=5$, so the range of the denominator is $[1, 5]$.

Hence the range of $f(x)$ is $\left[\dfrac{5}{5}, \dfrac{5}{1}\right] = [1, 5]$.

> **COMMENT**
>
> Given the denominator is always strictly positive, the minimum of $f(x)$ occurs when the denominator is at a maximum, and the maximum of $f(x)$ occurs when the denominator is at a minimum.

9 Trigonometric equations and identities

Exercise 9A

12 $2\sin x + 1 = 0$

$\sin x = -\dfrac{1}{2}$

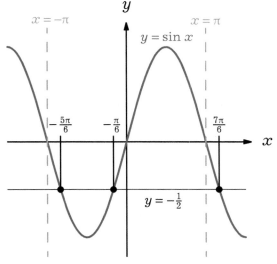

Figure 9A.12 Solutions to $\sin x = -\dfrac{1}{2}$ in $]-\pi, \pi[$

$\sin x = -\dfrac{1}{2}$ has the 2 solutions

$x_1 = \sin^{-1}\left(-\dfrac{1}{2}\right) = -\dfrac{\pi}{6}$ and $x_2 = \pi - x_1 = \dfrac{7\pi}{6}$

But x_2 is outside the interval $(-\pi, \pi)$, so subtract 2π: $\dfrac{7\pi}{6} - 2\pi = -\dfrac{5\pi}{6}$

$\therefore$ the solutions are $x = -\dfrac{5\pi}{6}, -\dfrac{\pi}{6}$

13 For example, if $x = 1$:

$\tan^{-1} x = \dfrac{\pi}{4}$

$\sin^{-1} x = \dfrac{\pi}{2}$

$\cos^{-1} x = 0$

Clearly in this case $\tan^{-1} x \neq \dfrac{\sin^{-1} x}{\cos^{-1} x}$

COMMENT

The false idea being disproved by counter-example here is that division 'passes through' the inversion of a function, i.e. that for a function $h(x) = \dfrac{f(x)}{g(x)}$ it should follow that

$h^{-1}(x) = \dfrac{f^{-1}(x)}{g^{-1}(x)}$; this is generally not true!

Exercise 9B

6 $3\cos x = \tan x$

$3\cos x = \dfrac{\sin x}{\cos x}$

$3\cos^2 x - \sin x = 0$

$3 - 3\sin^2 x - \sin x = 0$

$3\sin^2 x + \sin x - 3 = 0$

$\sin x = \dfrac{-1 \pm \sqrt{1+36}}{6}$

$\sin x = 0.847$ or -1.18 (reject as < -1)

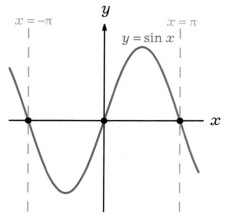

Figure 9B.6 Solutions to $\sin x = 0.847$ in $[0, 2\pi]$

There are 2 solutions to $\sin x = 0.847$ in $[0, 2\pi]$:

$x_1 = \sin^{-1} 0.847 = 1.01$

$x_2 = \pi - 1.01 = 2.13$

$\therefore x = 1.01, 2.13$

7 **a** $2\sin^2 x - 3\sin x = 2$

$2\sin^2 x - 3\sin x - 2 = 0$

$(2\sin x + 1)(\sin x - 2) = 0$

$\sin x = -\dfrac{1}{2}$ or 2 (reject as > 1)

$\therefore \sin x = -\dfrac{1}{2}$

b

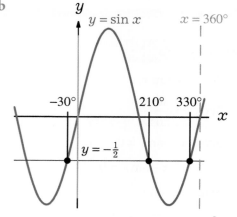

Figure 9B.7 Solutions to $\sin x = -\dfrac{1}{2}$ in $]0, 360°[$

$\sin x = -\dfrac{1}{2}$ has the 2 solutions

$x_1 = \sin^{-1}\left(-\dfrac{1}{2}\right) = -30°$ and

$x_2 = 180° - x_1 = 210°$

But x_1 is outside the interval $]0, 360°[$, so add $360°$: $-30° + 360° = 330°$

$\therefore$ the solutions are $x = 210°, 330°$

8 $\sin x \tan x = \sin^2 x$

$\dfrac{\sin^2 x}{\cos x} = \sin^2 x$

$\sin^2 x = \sin^2 x \cos x$

$\sin^2 x (\cos x - 1) = 0$

$\sin x = 0$ or $\cos x = 1$

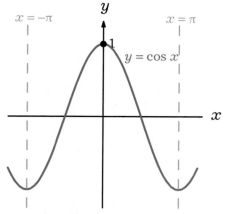

Figure 9B.8.1 Solutions to $\sin x = 0$ in $[-\pi, \pi]$

$\sin x = 0$

$\Rightarrow x = -\pi, 0, \pi$

Figure 9B.8.2 Solutions to $\cos x = 1$ in $[-\pi, \pi]$

$\cos x = 1 \Rightarrow x = 0$

∴ the solutions are $x = -\pi, 0, \pi$

9 $x \in]-\pi, \pi[\Rightarrow x^2 \in [0, \pi^2[$

To solve $\sin(x^2) = \dfrac{1}{2}$, first consider

$\sin \theta = \dfrac{1}{2}$:

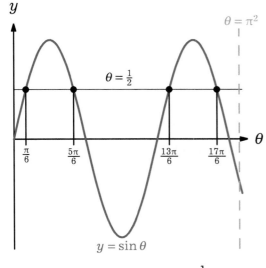

Figure 9B.9 Solutions to $\sin \theta = \dfrac{1}{2}$ in $[0, \pi^2[$

The first 2 solutions for $\theta = x^2$ are

$x_1^2 = \dfrac{\pi}{6}$

$x_2^2 = \pi - x_1^2 = \dfrac{5\pi}{6}$

and further periodic solutions are found by adding the period 2π:

$x^2 = \dfrac{\pi}{6}, \dfrac{5\pi}{6}, \dfrac{13\pi}{6}, \dfrac{17\pi}{6}$

∴ $x = \pm\sqrt{\dfrac{\pi}{6}}, \pm\sqrt{\dfrac{5\pi}{6}}, \pm\sqrt{\dfrac{13\pi}{6}}, \pm\sqrt{\dfrac{17\pi}{6}}$

Exercise 9C

7 a $3 - 2\tan^2 x = 3 - \dfrac{2\sin^2 x}{\cos^2 x}$

$= \dfrac{3\cos^2 x - 2\sin^2 x}{\cos^2 x}$

$= \dfrac{3\cos^2 x - 2(1 - \cos^2 x)}{\cos^2 x}$

$= \dfrac{5\cos^2 x - 2}{\cos^2 x}$

$= 5 - \dfrac{2}{\cos^2 x}$

b $\dfrac{1 + \tan^2 x}{\cos^2 x} = \dfrac{1 + \dfrac{\sin^2 x}{\cos^2 x}}{\cos^2 x}$

$= \dfrac{\cos^2 x + \sin^2 x}{(\cos^2 x)^2}$

$= \dfrac{1}{(1 - \sin^2 x)^2}$

8 $\sin^2 x + \cos^2 x = 1$

$\Rightarrow \dfrac{\sin^2 x}{\cos^2 x} + \dfrac{\cos^2 x}{\cos^2 x} = \dfrac{1}{\cos^2 x}$

$\Rightarrow \tan^2 x + 1 = \dfrac{1}{\cos^2 x}$

a From the above,

$\dfrac{1}{\cos^2 x} = 1 + t^2$

$\Rightarrow \cos^2 x = \dfrac{1}{1 + t^2}$

b $\sin^2 x = 1 - \cos^2 x$

$= 1 - \dfrac{1}{1 + t^2}$

$= \dfrac{1 + t^2 - 1}{1 + t^2}$

$= \dfrac{t^2}{1 + t^2}$

c $\cos^2 x - \sin^2 x = \dfrac{1}{1 + t^2} - \dfrac{t^2}{1 + t^2}$

$= \dfrac{1 - t^2}{1 + t^2}$

9 Trigonometric equations and identities

d $\dfrac{2}{\sin^2 x} + 1 = \dfrac{2}{\left(\dfrac{t^2}{1+t^2}\right)} + 1$

$= \dfrac{2(1+t^2)}{t^2} + 1$

$= 3 + \dfrac{2}{t^2}$

Exercise 9D

> **COMMENT**
>
> Always check solutions to ensure that values lie within the required interval, especially if the rearrangement has involved any division or multiplication. Sketching the graph is a good way of finding out how many solutions you need.

5 $5\sin^2 \theta = 4\cos^2 \theta$

$5\sin^2 \theta = 4(1 - \sin^2 \theta)$

$9\sin^2 \theta = 4$

$\sin^2 \theta = \dfrac{4}{9}$

$\sin \theta = \pm \dfrac{2}{3}$

There are 2 solutions to each of $\sin \theta = \dfrac{2}{3}$ and $\sin \theta = -\dfrac{2}{3}$ (positive and negative):

$\theta_1 = \sin^{-1}\left(\pm \dfrac{2}{3}\right) = \pm 41.8°$

$\theta_2 = 180° - \theta_1 = 138.2°, 221.8°$

But 221.8° is outside the interval $-180° \leq \theta \leq 180°$, so subtract 360°:

$221.8° - 360° = -138.2°$

∴ the solutions are $\theta = \pm 41.8°$, $\pm 138°$ (3SF)

6 $2\cos^2 t - \sin t - 1 = 0$

$2(1 - \sin^2 t) - \sin t - 1 = 0$

$2\sin^2 t + \sin t - 1 = 0$

$(2\sin t - 1)(\sin t + 1) = 0$

$\sin t = \dfrac{1}{2}$ or -1

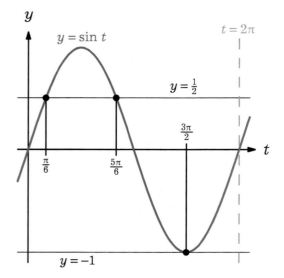

Figure 9D.6 Solutions to $\sin t = \dfrac{1}{2}$ and $\sin t = -1$ in $[0, 2\pi]$

There are 3 solutions in total.

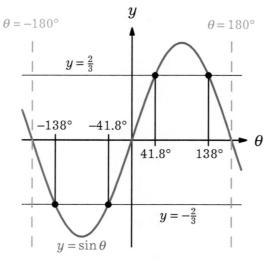

Figure 9D.5 Solutions to $\sin \theta = \pm \dfrac{2}{3}$ in $[-180°, 180°]$

78 Topic 9D Using identities to solve equations

For $\sin t = \frac{1}{2}$:

$t_1 = \sin^{-1}\left(\frac{1}{2}\right) = \frac{\pi}{6}$

$t_2 = \pi - t_1 = \frac{5\pi}{6}$

For $\sin t = -1$:

$t_1 = \sin^{-1}(-1) = -\frac{\pi}{2}$

But this is outside the interval $0 \le t \le 2\pi$, so add 2π: $-\frac{\pi}{2} + 2\pi = \frac{3\pi}{2}$

∴ the solutions are $t = \frac{\pi}{6}, \frac{5\pi}{6}, \frac{3\pi}{2}$

7 $4\cos^2 x - 5\sin x - 5 = 0$

$4(1 - \sin^2 x) - 5\sin x - 5 = 0$

$4\sin^2 x + 5\sin x + 1 = 0$

$(4\sin x + 1)(\sin x + 1) = 0$

$\sin x = -\frac{1}{4}$ or -1

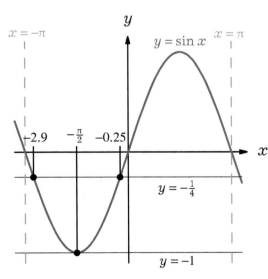

Figure 9D.7 Solutions to $\sin x = -\frac{1}{4}$ and $\sin x = -1$ in $[-\pi, \pi]$

There are 3 solutions in total.

For $\sin x = -\frac{1}{4}$:

$x_1 = \sin^{-1}\left(-\frac{1}{4}\right) = -0.253$

$x_2 = -\pi - x_1 = -2.89$

For $\sin x = -1$:

$x_1 = \sin^{-1}(-1) = -\frac{\pi}{2}$

∴ the solutions are $x = -2.89, -0.253, -\frac{\pi}{2}$

8 $\cos^2 t + 5\cos t = 2\sin^2 t$

$\cos^2 t + 5\cos t = 2(1 - \cos^2 t)$

$3\cos^2 t + 5\cos t - 2 = 0$

$(3\cos t - 1)(\cos t + 2) = 0$

$\cos t = \frac{1}{3}$ or -2 (reject as <-1)

∴ $\cos t = \frac{1}{3}$

9 a $6\sin^2 x + \cos x = 4$

$6(1 - \cos^2 x) + \cos x - 4 = 0$

$6\cos^2 x - \cos x - 2 = 0$

$(2\cos x + 1)(3\cos x - 2) = 0$

$\cos x = -\frac{1}{2}$ or $\frac{2}{3}$

b

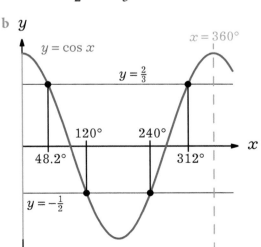

Figure 9D.9 Solutions to $\cos x = -\frac{1}{2}$ and $\cos x = \frac{2}{3}$ in $[0°, 360°]$

9 Trigonometric equations and identities

There are 2 solutions to each.

For $\cos x = -\dfrac{1}{2}$:

$x_1 = \cos^{-1}\left(-\dfrac{1}{2}\right) = 120°$

$x_2 = 360° - x_1 = 240°$

For $\cos x = \dfrac{2}{3}$:

$x_1 = \cos^{-1}\left(\dfrac{2}{3}\right) = 48.2°$

$x_2 = 360° - x_1 = 312°$

∴ the solutions are
$x = 48.2°, 120°, 240°, 312°$

10 a $2\sin^2 x - 3\sin x \cos x + \cos^2 x = 0$

$\dfrac{2\sin^2 x}{\cos^2 x} - \dfrac{3\sin x \cos x}{\cos^2 x} + \dfrac{\cos^2 x}{\cos^2 x} = 0$

$2\tan^2 x - 3\tan x + 1 = 0$

b $2\tan^2 x - 3\tan x + 1 = 0$

$(2\tan x - 1)(\tan x - 1) = 0$

$\tan x = \dfrac{1}{2}$ or 1

There are 2 solutions to each.

For $\tan x = \dfrac{1}{2}$:

$x_1 = \tan^{-1}\left(\dfrac{1}{2}\right) = 0.464$

$x_2 = x_1 - \pi = -2.68$

For $\tan x = 1$:

$x_1 = \tan^{-1}(1) = \dfrac{\pi}{4}$

$x_2 = x_1 - \pi = -\dfrac{3\pi}{4}$

∴ the solutions are
$x = -2.68, -\dfrac{3\pi}{4}, 0.464, \dfrac{\pi}{4}$

Exercise 9E

6 $\cos^2 \theta + \cos 2\theta = 0$

$\cos^2 \theta + 2\cos^2 \theta - 1 = 0$

$3\cos^2 \theta = 1$

$\cos^2 \theta = \dfrac{1}{3}$

$\cos \theta = \pm \dfrac{1}{\sqrt{3}}$

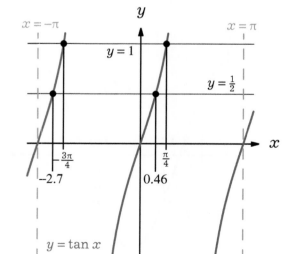

Figure 9D.10 Solutions to $\tan x = \dfrac{1}{2}$ and $\tan x = 1$ in $]-\pi, \pi[$

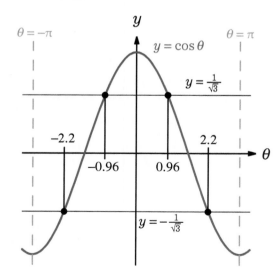

Figure 9E.6 Solutions to $\cos x = \dfrac{1}{\sqrt{3}}$ and $\cos x = -\dfrac{1}{\sqrt{3}}$ in $[-\pi, \pi]$

There are 2 solutions to each (positive and negative):

$\theta_1 = \cos^{-1}\left(\pm\dfrac{1}{\sqrt{3}}\right) = 0.955, 2.19$

$\theta_2 = -\theta_1 = -0.955, -2.19$

∴ the solutions are $\theta = \pm 0.955, \pm 2.19$

7 $\dfrac{1-\cos 2\theta}{1+\cos 2\theta} = \dfrac{1-(2\cos^2\theta - 1)}{1+(2\cos^2\theta - 1)}$

$= \dfrac{2-2\cos^2\theta}{2\cos^2\theta}$

$= \dfrac{2(1-\cos^2\theta)}{2\cos^2\theta}$

$= \dfrac{2\sin^2\theta}{2\cos^2\theta}$

$= \tan^2\theta$

8 a $\cos 4\theta = 2\cos^2 2\theta - 1$

$= 2(2\cos^2\theta - 1)^2 - 1$

$= 2(4\cos^4\theta - 4\cos^2\theta + 1) - 1$

$= 8\cos^4\theta - 8\cos^2\theta + 1$

b

$\cos 4\theta = 8\cos^4\theta - 8\cos^2\theta + 1$

$= 8(1-\sin^2\theta)^2 - 8(1-\sin^2\theta) + 1$

$= 8(1 - 2\sin^2\theta + \sin^4\theta) - 8(1-\sin^2\theta) + 1$

$= 8 - 16\sin^2\theta + 8\sin^4\theta - 8 + 8\sin^2\theta + 1$

$= 8\sin^4\theta - 8\sin^2\theta + 1$

9 a i $\cos^2\left(\dfrac{x}{2}\right) = \dfrac{1}{2}\left(2\cos^2\left(\dfrac{x}{2}\right)\right)$

$= \dfrac{1}{2}\left(\cos\left(2\times\dfrac{x}{2}\right) + 1\right)$

$= \dfrac{1}{2}(\cos x + 1)$

ii $\sin^2\left(\dfrac{x}{2}\right) = \dfrac{1}{2}\left(2\sin^2\left(\dfrac{x}{2}\right)\right)$

$= \dfrac{1}{2}\left(1 - \cos\left(2\times\dfrac{x}{2}\right)\right)$

$= \dfrac{1}{2}(1 - \cos x)$

b $\tan^2\left(\dfrac{x}{2}\right) = \dfrac{\sin^2\left(\dfrac{x}{2}\right)}{\cos^2\left(\dfrac{x}{2}\right)} = \dfrac{1-\cos x}{1+\cos x}$

10 $a\sin 4x = b\sin 2x$

$2a\sin 2x \cos 2x = b\sin 2x$

$2a\sin 2x \cos 2x - b\sin 2x = 0$

$\sin 2x(2a\cos 2x - b) = 0$

$\sin 2x = 0 \quad \text{or} \quad \cos 2x = \dfrac{b}{2a}$

But $\sin 2x \neq 0$ for $0 < x < \dfrac{\pi}{2}$

∴ $\cos 2x = \dfrac{b}{2a}$

$1 - 2\sin^2 x = \dfrac{b}{2a}$

$\sin^2 x = \dfrac{1 - \dfrac{b}{2a}}{2} = \dfrac{2a-b}{4a}$

Mixed examination practice 9

Short questions

1

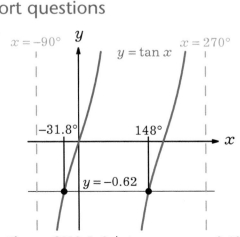

Figure 9MS.1 Solutions to $\tan x = -0.62$ in $]-90°, 270°[$

There are 2 solutions:

$x_1 = \tan^{-1}(-0.62) = -31.8°$

$x_2 = x_1 + \pi = 148°$

$\therefore x = -31.8°, 148°$

2 a $\cos\theta = \sqrt{1-\sin^2\theta}$

$= \sqrt{1-\left(\dfrac{2}{3}\right)^2}$

$= \sqrt{\dfrac{9-4}{9}} = \dfrac{\sqrt{5}}{3}$

(Select the positive root as $0 < \theta < \dfrac{\pi}{2}$.)

b $\cos(2\theta) = 1 - 2\sin^2(\theta)$

$= 1 - 2\left(\dfrac{2}{3}\right)^2$

$= \dfrac{1}{9}$

3 $5\sin^2\theta = 4\cos^2\theta$

$5\sin^2\theta = 4(1-\sin^2\theta)$

$9\sin^2\theta = 4$

$\sin^2\theta = \dfrac{4}{9}$

$\sin\theta = \pm\dfrac{2}{3}$

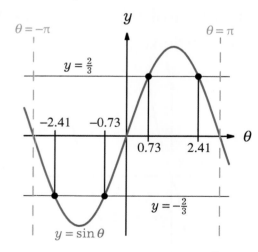

Figure 9MS.3 Solutions to $\sin\theta = \dfrac{2}{3}$ and $\sin\theta = -\dfrac{2}{3}$ in $[-\pi, \pi]$

There are 2 solutions to each (positive and negative):

$\theta_1 = \sin^{-1}\left(\pm\dfrac{2}{3}\right) = \pm 0.730$

$\theta_2 = \pi - \theta_1 = 2.41, 3.87$

But 3.87 is outside the interval $-\pi \le \theta \le \pi$, so subtract 2π: $3.87 - 2\pi = -2.41$

$\therefore \theta = \pm 0.730, \pm 2.41$

4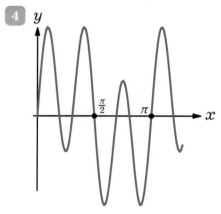

Figure 9MS.4 Graph of $y = \sin(2x) + 2\sin(6x)$

From the graph, the wave-form repeats every π units in the x direction, so the period is π.

5 $\dfrac{2}{\cos^2 x} - \tan^2 x = \dfrac{2}{\cos^2 x} - \dfrac{\sin^2 x}{\cos^2 x}$

$= \dfrac{2-\sin^2 x}{\cos^2 x}$

$= \dfrac{2-2\sin^2 x + \sin^2 x}{\cos^2 x}$

$= \dfrac{2(1-\sin^2 x) + \sin^2 x}{\cos^2 x}$

$= \dfrac{2\cos^2 x + \sin^2 x}{\cos^2 x}$

$= 2 + \dfrac{\sin^2 x}{\cos^2 x}$

$= 2 + \tan^2 x$

82 Mixed examination practice 9

6 $\cos\theta - 2\sin^2\theta + 2 = 0$

$\cos\theta - 2(1-\cos^2\theta) + 2 = 0$

$\cos\theta - 2 + 2\cos^2\theta + 2 = 0$

$\cos\theta(1 + 2\cos\theta) = 0$

$\cos\theta = 0 \quad \text{or} \quad -\dfrac{1}{2}$

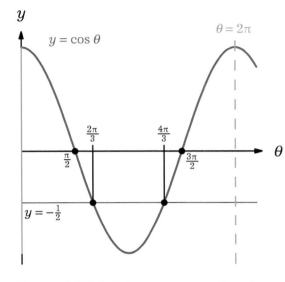

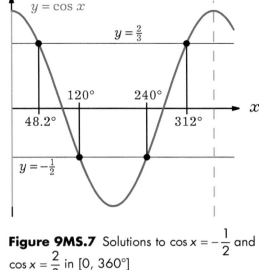

Figure 9MS.6 Solutions to $\cos\theta = 0$ and $\cos\theta = -\dfrac{1}{2}$ in $[0, 2\pi]$

There are 2 solutions to each:

$\cos\theta = 0$

$\Rightarrow \theta = \dfrac{\pi}{2}, \dfrac{3\pi}{2}$

$\cos\theta = -\dfrac{1}{2}$

$\Rightarrow \theta = \dfrac{2\pi}{3}, \dfrac{4\pi}{3}$

$\therefore \theta = \dfrac{\pi}{2}, \dfrac{3\pi}{2}, \dfrac{2\pi}{3}, \dfrac{4\pi}{3}$

7 $6\sin^2 x + \cos x = 4$

$6(1-\cos^2 x) + \cos x = 4$

$6\cos^2 x - \cos x - 2 = 0$

$(2\cos x + 1)(3\cos x - 2) = 0$

$\cos x = -\dfrac{1}{2} \quad \text{or} \quad \dfrac{2}{3}$

Figure 9MS.7 Solutions to $\cos x = -\dfrac{1}{2}$ and $\cos x = \dfrac{2}{3}$ in $[0, 360°]$

There are 2 solutions to each.

For $\cos x = -\dfrac{1}{2}$:

$x_1 = \cos^{-1}\left(-\dfrac{1}{2}\right) = 120°$

$x_2 = 360° - 120° = 240°$

For $\cos x = \dfrac{2}{3}$:

$x_1 = \cos^{-1}\left(\dfrac{2}{3}\right) = 48.2°$

$x_2 = 360° - 48.2° = 312°$

$\therefore x = 48.2°, 120°, 240°, 312°$

8 $\sin 2\theta = \cos\theta$

$2\sin\theta\cos\theta = \cos\theta$

$\cos\theta(2\sin\theta - 1) = 0$

$\cos\theta = 0 \quad \text{or} \quad \sin\theta = \dfrac{1}{2}$

For $0 \leq \theta \leq 2\pi$ there are 2 solutions to each:

$\cos\theta = 0$

$\Rightarrow \theta = \dfrac{\pi}{2}, \dfrac{3\pi}{2}$

$\sin\theta = \dfrac{1}{2}$

$\Rightarrow \theta = \dfrac{\pi}{6}, \dfrac{5\pi}{6}$

$\therefore \theta = \dfrac{\pi}{2}, \dfrac{3\pi}{2}, \dfrac{\pi}{6}, \dfrac{5\pi}{6}$

Long questions

1

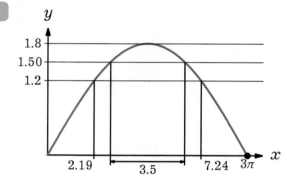

Figure 9ML.1 Graph of $y = 1.8\sin\left(\dfrac{x}{3}\right)$ between $x = 0$ and the first positive zero

a The width of the river is the x-coordinate of the first positive zero: $3\pi = 9.42$ metres (3SF)

b The maximum width of the barge is the distance between the first 2 positive solutions of $1.8\sin\left(\dfrac{x}{3}\right) = 1.2$.

From GDC, the solutions are $x = 2.19$ and $x = 7.24$, so the maximum width of the barge is $7.24 - 2.19 = 5.05$ metres.

c The centre of the bridge is at $x = \dfrac{3\pi}{2} = 4.71$ metres.

This barge of width 2.5 m, travelling along the centre of the river course, will be positioned in the interval $[4.71 - 1.75, 4.71 + 1.75] = [2.96, 6.46]$

In this interval, $y \geq 1.8\sin\left(\dfrac{2.96}{3}\right) = 1.50$, so the maximum height of the barge is 1.50 metres above water level.

2 a From GDC:

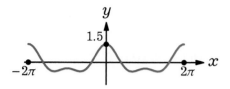

Figure 9ML.2 Graph of $C(x) = \cos x + \dfrac{1}{2}\cos 2x$ for $-2\pi \leq x \leq 2\pi$

b Since $\cos(a + 2k\pi) = \cos a$ for any integer k,

$C(x + 2\pi) = \cos(x + 2\pi) + \dfrac{1}{2}\cos(2(x + 2\pi))$

$= \cos(x + 2\pi) + \dfrac{1}{2}\cos(2x + 4\pi)$

$= \cos x + \dfrac{1}{2}\cos 2x$

$= C(x)$

So $C(x)$ is periodic with period a factor of 2π.

From Figure 9ML.2 it is clear that the period is no less than 2π, so the period equals 2π.

c From GDC, $C(x)$ has maximum points at $x = 0, \pm\pi, \pm2\pi$

d $\cos x + \dfrac{1}{2}\cos 2x = 0$

$\cos x + \dfrac{1}{2}(2\cos^2 x - 1) = 0$

$\cos^2 x + \cos x - \dfrac{1}{2} = 0$

$\cos x = \dfrac{-1 \pm \sqrt{1^2 - 4(1)\left(-\dfrac{1}{2}\right)}}{2}$

$= -\dfrac{1}{2} \pm \dfrac{\sqrt{3}}{2}$

The smallest positive root is given by

$x_0 = \cos^{-1}\left(-\dfrac{1}{2} + \dfrac{\sqrt{3}}{2}\right) = 1.2$ (2SF).

e i $\cos x = \cos(-x)$ for all x

$$\therefore C(-x) = \cos(-x) + \frac{1}{2}\cos(-2x)$$
$$= \cos x + \frac{1}{2}\cos 2x$$
$$= C(x)$$

ii Using parts (e)(i) and (b):

$$C(x) = C(-x)$$
$$= C(2\pi - x)$$
$$\therefore x_1 = 2\pi - x_0$$

3 a Repeated root $\Rightarrow$ discriminant $= 0$:

$$\Delta = k^2 - 16 = 0$$
$$\Rightarrow k = \pm 4$$

b $4\sin^2\theta = 5 - k\cos\theta$

$$4(1 - \cos^2\theta) = 5 - k\cos\theta$$
$$4\cos^2 x - k\cos x + 1 = 0$$

c i $f_4(\theta) = 4\cos^2\theta - 4\cos\theta + 1$

From (a), $4x^2 - kx + 1 = 0$ has a repeated root, so there is a single value of $\cos\theta$ which satisfies $f_4(\theta) = 0$.

ii $f_4(\theta) = 0$

$$4\cos^2\theta - 4\cos\theta + 1 = 0$$
$$(2\cos\theta - 1)^2 = 0$$
$$\cos\theta = \frac{1}{2}$$

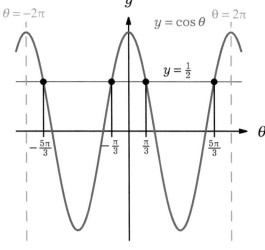

Figure 9ML.3.1 Solutions to $\cos\theta = \frac{1}{2}$ in $[-2\pi, 2\pi]$

There are 4 solutions. The first 2 are:

$$\theta_1 = \cos^{-1}\left(\frac{1}{2}\right) = \frac{\pi}{3}$$

$$\theta_2 = -\theta_1 = -\frac{\pi}{3}$$

Then, adding/subtracting 2π gives

$$\frac{\pi}{3} - 2\pi = -\frac{5\pi}{3} \quad \text{and} \quad -\frac{\pi}{3} + 2\pi = \frac{5\pi}{3}$$

$$\therefore \theta = \pm\frac{\pi}{3}, \pm\frac{5\pi}{3}$$

iii Substituting $x = 1$ into $4x^2 - kx + 1 = 0$:

$$4 - k + 1 = 0$$
$$k = 5$$

iv With $k = 5$,

$$f_5(\theta) = 0$$
$$4\cos^2\theta - 5\cos\theta + 1 = 0$$
$$(4\cos\theta - 1)(\cos\theta - 1) = 0$$
$$\cos\theta = \frac{1}{4} \quad \text{or} \quad 1$$

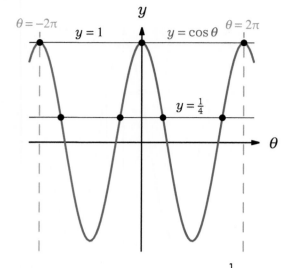

Figure 9ML.3.2 Solutions to $\cos\theta = \frac{1}{4}$ and $\cos\theta = 1$ in $[-2\pi, 2\pi]$

From the graph,

$\cos\theta = \frac{1}{4}$ has 4 solutions in $[-2\pi, 2\pi]$ and

$\cos\theta = 1$ has 3 solutions.

In total, there are 7 solutions in $[-2\pi, 2\pi]$.

10 Geometry of triangles and circles

COMMENT

In questions on geometric shapes, if no diagram is given it is usually wise to draw a quick sketch. This reduces the opportunity for errors of interpretation and makes it easier to check for the sense of an answer.
When checking for sense, always remember that in a triangle, the widest angle lies opposite the longest side and the narrowest angle lies opposite the shortest side.

Exercise 10B

4

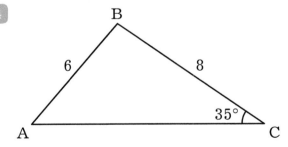

Figure 10B.4

By the sine rule:

$$\frac{\sin C\hat{A}B}{BC} = \frac{\sin A\hat{C}B}{AB}$$

$$\sin C\hat{A}B = \frac{BC \sin A\hat{C}B}{AB} = \frac{8 \sin 35°}{6}$$

Two solutions for $C\hat{A}B$ are

$$x_1 = \sin^{-1}\left(\frac{8 \sin 35°}{6}\right) = 49.9°$$

and $x_2 = 180° - 49.9° = 130.1°$

and hence $A\hat{B}C = 180° - 49.9° - 35° = 95.1°$
or $180° - 130.1° - 35° = 14.9°$ (both solutions are viable).

To find AC, use the sine rule:

$$\frac{\sin A\hat{C}B}{AB} = \frac{\sin A\hat{B}C}{AC}$$

$$AC = \frac{AB \sin A\hat{B}C}{\sin A\hat{C}B} = \frac{6 \sin A\hat{B}C}{\sin 35°}$$

∴ two possible triangles exist:

one with angles 35°, 49.9°, 95.1° and AC = 10.4 cm;

another with angles 35°, 130°, 14.9° and AC = 2.69 cm.

5 By sine rule in triangle ABD:

$$\frac{\sin A\hat{B}D}{AD} = \frac{\sin A\hat{D}B}{AB}$$

$$A\hat{B}D = \sin^{-1}\left(\frac{AD \sin A\hat{D}B}{AB}\right)$$

$$= \sin^{-1}\left(\frac{5 \sin 75°}{6}\right)$$

$$= 53.6°$$

By sine rule in triangle ABC:

$$\frac{\sin A\hat{C}B}{AB} = \frac{\sin A\hat{B}C}{AC}$$

$$A\hat{C}B = \sin^{-1}\left(\frac{AB \sin A\hat{B}C}{AC}\right)$$

$$= \sin^{-1}\left(\frac{6\sin 53.6°}{8}\right)$$

$$= 37.1°$$

Then $B\hat{A}C = 180° - 53.6° - 37.1° = 89.3°$

By sine rule in triangle ABC:

$$\frac{BC}{\sin B\hat{A}C} = \frac{AC}{\sin A\hat{B}C}$$

$$BC = \frac{AC \sin B\hat{A}C}{\sin A\hat{B}C}$$

$$= \frac{8\sin 89.3°}{\sin 53.6°}$$

$$= 9.94 \text{ cm}$$

6 The question has been amended to state that the peg lies due south of the balloon and the observer stands due north of the balloon.

Let the observer be at O, the balloon at B and the peg at P. Let the point K lie directly below O such that PK is horizontal, and let L lie above O such that LB is horizontal.

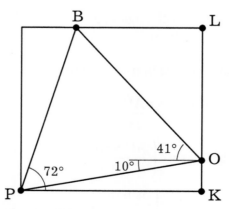

Figure 10B.6

$K\hat{P}B = 72°$, $K\hat{P}O = 10°$
and $P\hat{O}B = 10° + 41° = 51°$

Also,

$O\hat{P}B = 72° - 10° = 62°$

∴ $P\hat{B}O = 180° - 62° - 51° = 67°$

By sine rule in triangle POB,

$$\frac{OP}{\sin O\hat{B}P} = \frac{BP}{\sin P\hat{O}B}$$

$$OP = \frac{BP \sin O\hat{B}P}{\sin P\hat{O}B}$$

$$= \frac{20\sin 67°}{\sin 51°}$$

$$= 23.7 \text{ m}$$

Then, by trigonometry in triangle POK,

$PK = OP\cos 10° = 23.3 \text{ m}$

7 By the sine rule:

$$\frac{\sin A\hat{C}B}{AB} = \frac{\sin A\hat{B}C}{AC}$$

$$\sin A\hat{C}B = \frac{AB\sin A\hat{B}C}{AC}$$

$$= \frac{12\sin 47°}{8}$$

$$= 1.097$$

But since $\sin x < 1$ for any angle x in a triangle, this is not possible.

Exercise 10C

4

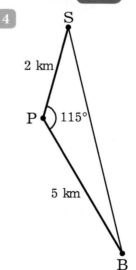

Figure 10C.4

$B\hat{P}S = 130° - 15° = 115°$

By the cosine rule:

$BS = \sqrt{(BP)^2 + (PS)^2 - (BP)(PS)\cos B\hat{P}S}$

$= \sqrt{5^2 + 2^2 - 2(5)(2)\cos 115°}$

$= 6.12$ km

5 By cosine rule in triangle ACD:

$A\hat{D}C = \cos^{-1}\left(\dfrac{(AD)^2 + (CD)^2 - (AC)^2}{2(AD)(CD)}\right)$

$= \cos^{-1}\left(\dfrac{6^2 + 7^2 - 10^2}{2(6)(7)}\right)$

$= 100.3°$

$\therefore B\hat{D}C = 180° - A\hat{D}C = 79.7°$

By sine rule in triangle BCD:

$\dfrac{BC}{\sin B\hat{D}C} = \dfrac{DC}{\sin D\hat{B}C}$

$BC = \dfrac{DC \sin B\hat{D}C}{\sin D\hat{B}C}$

$= \dfrac{7 \sin 79.7°}{\sin 60°}$

$= 7.95$

$\therefore x = 7.95$

6

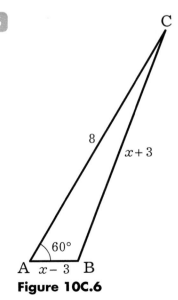

Figure 10C.6

By the cosine rule,

$BC^2 = AB^2 + AC^2 - 2(AB)(AC)\cos B\hat{A}C$

$(x+3)^2 = (x-3)^2 + 8^2 - 2 \times 8(x-3) \times \dfrac{1}{2}$

$x^2 + 6x + 9 = x^2 - 6x + 9 + 64 - 8x + 24$

$20x = 88$

$x = 4.4$

7

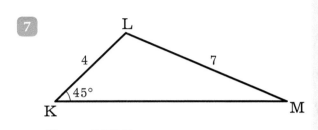

Figure 10C.7

By the cosine rule,

$(LM)^2 = (KL)^2 + (KM)^2 - 2(KL)(KM)\cos L\hat{K}M$

$\therefore 7^2 = 4^2 + x^2 - 2 \times 4x \times \dfrac{1}{\sqrt{2}}$

$49 = 16 + x^2 - \dfrac{8x}{\sqrt{2}}$

$x^2 - 4\sqrt{2}x - 33 = 0$

$x = \dfrac{4\sqrt{2} \pm \sqrt{32 - 4 \times 1 \times (-33)}}{2}$

$= 2\sqrt{2} \pm \sqrt{41}$

Since the length must be positive,

$KM = 2\sqrt{2} + \sqrt{41}$

Exercise 10D

3

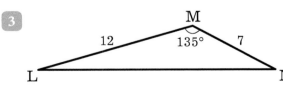

Figure 10D.3

10 Geometry of triangles and circles 89

By the cosine rule,

$$LN = \sqrt{(LM)^2 + (MN)^2 - 2(LM)(MN)\cos L\hat{M}N}$$
$$= \sqrt{12^2 + 7^2 - 2(12)(7)\cos 135°}$$
$$= 17.7 \text{ cm}^2$$

$$\text{Area} = \frac{1}{2}(LM)(MN)\sin L\hat{M}N$$
$$= \frac{1}{2} \times 12 \times 7 \sin 135°$$
$$= 29.7 \text{ cm}^2$$

4

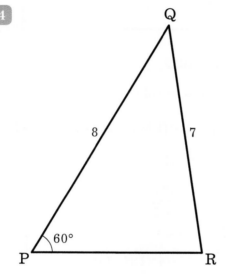

Figure 10D.4

Let $PR = x$. By the cosine rule,

$$(RQ)^2 = (PQ)^2 + (PR)^2 - 2(PQ)(PR)\cos R\hat{P}Q$$
$$7^2 = 8^2 + x^2 - 2 \times 8x \cos 60°$$
$$49 = 64 + x^2 - 8x$$
$$x^2 - 8x + 15 = 0$$
$$(x-3)(x-5) = 0$$
$$x = 3 \text{ or } 5$$

Area of triangle $= \frac{1}{2}(PQ)x \sin R\hat{P}Q$,

where x can take the two possible values above.

Area difference $= \frac{1}{2}(PQ)\sin R\hat{P}Q \times (x_1 - x_2)$

$$= \frac{1}{2} \times 8 \times \frac{\sqrt{3}}{2} \times (5-3)$$
$$= 4\sqrt{3} \text{ cm}^2$$

Exercise 10E

2

> **COMMENT**
>
> There could be several possible answers to this question, depending on which cuboid edge (if any) is included in ABC. If the question had specified that no cuboid edges are included in ABC, then the solution narrows down to case (iv), which is the answer given in the back of the coursebook.

Case (i):

If ABC includes a side of length 12.5 (right triangle), then the sides are 12.5, $\sqrt{10^2 + 7.3^2} = 12.4$ and $\sqrt{12.5^2 + 10^2 + 7.3^2} = 17.6$

From trigonometry, the angles are 90°, $\sin^{-1}\left(\frac{12.4}{17.6}\right) = 44.7°$ and 45.3°

$$\text{Area} = \frac{1}{2} \times 12.4 \times 12.5 = 77.4 \text{ cm}^2$$

Case (ii):

If ABC includes a side of length 10 (right triangle), then the sides are 10, $\sqrt{12.5^2 + 7.3^2} = 14.5$ and $\sqrt{12.5^2 + 10^2 + 7.3^2} = 17.6$

From trigonometry, the angles are 90°, $\sin^{-1}\left(\frac{10}{17.6}\right) = 34.6°$ and 55.4°

$$\text{Area} = \frac{1}{2} \times 10 \times 14.5 = 72.4 \text{ cm}^2$$

Case (iii):

If ABC includes a side of length 7.3 (right triangle), then the sides are 7.3,
$\sqrt{10^2 + 12.5^2} = 16.0$ and
$\sqrt{12.5^2 + 10^2 + 7.3^2} = 17.6$

From trigonometry, the angles are 90°,
$\sin^{-1}\left(\dfrac{7.3}{17.6}\right) = 24.5°$ and 65.5°

Area $= \dfrac{1}{2} \times 7.3 \times 16.0 = 58.4 \text{ cm}^2$

Case (iv):

If ABC includes no sides of the cuboid (oblique triangle), then the sides are

$\sqrt{10^2 + 7.3^2} = 12.4$,
$\sqrt{10^2 + 12.5^2} = 16.0$,
$\sqrt{12.5^2 + 7.3^2} = 14.5$

By applying the cosine rule repeatedly, the angles are 47.6°, 59.7° and 72.7°

(For example, the angle between the sides of lengths 12.4 and 16.0 is

$\cos^{-1}\left(\dfrac{12.4^2 + 16.0^2 - 14.5^2}{2(12.4)(16.0)}\right) = 59.7°$.)

Area $= \dfrac{1}{2} \times 12.4 \times 14.5 \times \cos(72.7°)$
$= 85.6 \text{ cm}^2$

3 By trigonometry, the flagpole height BF is $12 \tan 52° = 15.4$ m

$\therefore \tan \theta = \dfrac{BF}{8}$

$\theta = \tan^{-1}\left(\dfrac{BF}{8}\right) = 62.5°$

4

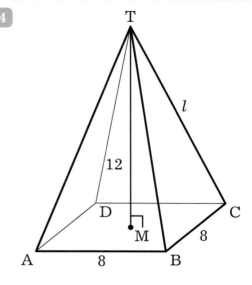

Figure 10E.4.1

Let M be the midpoint of the base. If one corner of the base is A and the apex of the pyramid is T, then by Pythagoras' Theorem,

$AM = \dfrac{\sqrt{8^2 + 8^2}}{2}$

$= \dfrac{8\sqrt{2}}{2}$

$= 4\sqrt{2}$

Figure 10E.4.2

∴ $AT = \sqrt{12^2 + (4\sqrt{2})^2}$
$= \sqrt{144 + 32}$
$= \sqrt{176}$
$= 13.3 \text{ cm}$

i.e. $l = 13.3 \text{ cm}$

Figure 10E.4.3

5 **a** By Pythagoras' Theorem,

$CM = \sqrt{17^2 - 12^2}$
$= \sqrt{145}$
$= 12.0 \text{ cm}$

b CMB is an isosceles right triangle with $C\hat{M}B = 90°$ (the diagonals of a square meet at a right angle).

∴ $CB = \sqrt{2(CM)^2}$
$= \sqrt{2} \, CM$
$= \sqrt{2}\sqrt{145}$
$= \sqrt{290}$
$= 17.0 \text{ cm}$

6 **a** $QP = AQ \tan Q\hat{A}P$
$= 25 \tan 37°$
$= 18.8$

∴ $h = 18.8 \text{ m}$

b $QB = \dfrac{QP}{\tan Q\hat{B}P}$
$= \dfrac{25 \tan 37°}{\tan 42°}$
$= 20.9$

By the cosine rule,

$AB = \sqrt{(QB)^2 + (QA)^2 - 2(QB)(QA)\cos A\hat{Q}B}$
$= \sqrt{20.9^2 + 25^2 - 2(20.9)(25)\cos 75°}$
$= 28.1 \text{ m}$

7 **a** $\tan \alpha = \dfrac{h}{RA}$ and $\tan \beta = \dfrac{h}{RB}$

∴ $RA = \dfrac{h}{\tan \alpha}$ and $RB = \dfrac{h}{\tan \beta}$

By Pythagoras' Theorem,

$(AB)^2 = (RA)^2 + (RB)^2$

i.e. $d^2 = \left(\dfrac{h}{\tan \alpha}\right)^2 + \left(\dfrac{h}{\tan \beta}\right)^2$

$= h^2 \left(\dfrac{1}{\tan^2 \alpha} + \dfrac{1}{\tan^2 \beta}\right)$

b $\alpha = 45° \Rightarrow \tan \alpha = 1 \Rightarrow \dfrac{1}{\tan^2 \alpha} = 1$

$\beta = 30° \Rightarrow \tan \beta = \dfrac{1}{\sqrt{3}} \Rightarrow \dfrac{1}{\tan^2 \beta} = 3$

$h^2 = \dfrac{d^2}{\left(\dfrac{1}{\tan^2 \alpha} + \dfrac{1}{\tan^2 \beta}\right)}$

$= \dfrac{26^2}{1 + 3}$

$= 169$

∴ $h = 13 \text{ m}$

Exercise 10F

3 $l = r\theta$
$= 10 \times 2.5$
$= 25$ cm

4 a $\theta = \dfrac{l}{r}$
$= \dfrac{7.5}{8}$
$= 0.9375$ radians

b 0.9375 radians $= 0.9375 \times \dfrac{180°}{\pi}$
$= 53.7°$

5 Let θ be the angle subtended by the major arc; then

$\theta = \dfrac{l}{r}$
$= \dfrac{15}{4}$
$= 3.75$

$\therefore \hat{\text{MCN}} = 2\pi - 3.75$
$= 2.53$ radians

6 $r = \dfrac{l}{\theta}$
$= \dfrac{12}{1.6}$
$= 7.5$ cm

7

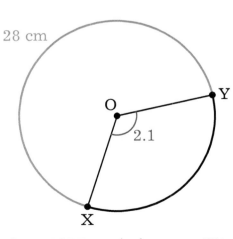

Figure 10F.7 Length of major arc XY is 28 cm

The angle subtended by the major arc is $2\pi - 2.1$, so

$r = \dfrac{l}{\theta}$
$= \dfrac{28}{2\pi - 2.1}$
$= 6.69$ cm

8 The perimeter p is composed of three arcs with radius 5 cm and angle $60° = \dfrac{\pi}{3}$

$\therefore p = 3\left(5 \times \dfrac{\pi}{3}\right)$
$= 5\pi = 15.7$ cm

9 Let l be the length of the arc; then

$l = r\theta$
$= 8 \times 0.7$
$= 5.6$

$\therefore p = 2(5+8) + 5.6$
$= 31.6$ cm

10 The angle between the two 5 cm sides is

$\theta = 180° - 2 \times 15°$
$= 150° = \dfrac{5\pi}{6}$

$\therefore p = 10 + 5 \times \dfrac{5\pi}{6}$
$= \left(10 + \dfrac{25\pi}{6}\right)$ cm

11

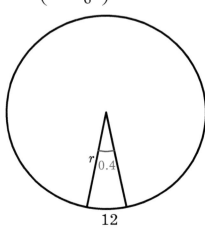

Figure 10F.11

The perimeter of the sector is made up of two radii and an arc:

$p = 2r + r\theta$

$12 = r(2 + 0.4)$

$r = \dfrac{12}{2.4} = 5$ cm

12 On the cone, the slant height is

$\sqrt{12^2 + \left(\dfrac{18}{2}\right)^2} = 15$, which is the radius r of the sector.

The perimeter of the base is 18π, which is the arc length l of the sector.

$\theta = \dfrac{l}{r}$

$= \dfrac{18\pi}{15}$

$= 1.2\pi$

$= 3.77$ radians (or $216°$)

Exercise 10G

3 $A = \dfrac{1}{2}r^2\theta$

$\Rightarrow \theta = \dfrac{2A}{r^2}$

$= \dfrac{2 \times 40}{10^2}$

$= 0.8$ radians

4 The larger angle is

$\theta = \dfrac{2A}{r^2}$

$= \dfrac{2 \times 744}{21^2}$

$= 3.37$

∴ smaller angle $= 2\pi - 3.37$

$= 2.91$ radians

(or $2.91 \times \dfrac{180°}{\pi} = 167°$)

5 $A = \dfrac{1}{2}r^2\theta$

$\Rightarrow r = \sqrt{\dfrac{2A}{\theta}}$

$= \sqrt{\dfrac{2 \times 54}{1.2}}$

$= \sqrt{90} = 9.49$ cm

6 $162° = 162 \times \dfrac{\pi}{180} = 2.827$ radians

$r = \sqrt{\dfrac{2A}{\theta}}$

$= \sqrt{\dfrac{2 \times 180}{2.827}}$

$= 11.3$ cm

7 $\theta = 45° = \dfrac{\pi}{4}$

Sector area $= \dfrac{r^2\theta}{2}$

$= \dfrac{6^2\pi}{8}$

$= 14.14$ cm^2

Triangle area $= \dfrac{1}{2} \times 6 \times 3 = 9$ cm^2

∴ shaded area $= 14.1 - 9 = 5.14$ cm^2

8 The perimeter of the sector is made up of two radii and an arc:

$p = 2r + r\theta$

$28 = r(2 + 1.6)$

$\Rightarrow r = \dfrac{28}{3.6} = 7.78$ cm

∴ $A = \dfrac{r^2\theta}{2} = \dfrac{1}{2}\left(\dfrac{28}{3.6}\right)^2(1.6) = 48.4$ cm^2

9 $p = 2r + r\theta$
$7 = r(2+\theta)$ ...(1)

$A = \dfrac{r^2\theta}{2}$

$3 = \dfrac{r^2\theta}{2}$

$\Rightarrow \theta = \dfrac{6}{r^2}$ ...(2)

Substituting (2) into (1):

$r\left(2 + \dfrac{6}{r^2}\right) = 7$

$2r^2 - 7r + 6 = 0$

$(2r-3)(r-2) = 0$

$r = 1.5$ or 2

So the radius is 1.5 cm or 2 cm.

10 Let θ be the minor sector angle. Then:

major sector area $A_1 = \dfrac{r^2(2\pi - \theta)}{2}$

minor sector area $A_2 = \dfrac{r^2\theta}{2}$

$A_1 - A_2 = 15$

$\dfrac{r^2(2\pi - \theta)}{2} - \dfrac{r^2\theta}{2} = 15$

$\dfrac{5^2}{2}(2\pi - 2\theta) = 15$

$\pi - \theta = \dfrac{15}{25}$

$\therefore \theta = \pi - 0.6 = 2.54$ radians

Exercise 10H

4 a Minor segment area $= \dfrac{r^2}{2}(\theta - \sin\theta)$

$= \dfrac{5^2}{2}(\theta - \sin\theta)$

$= 12.5(\theta - \sin\theta)$ cm^2

b $12.5(\theta - \sin\theta) = 15$

$\theta - \sin\theta = 1.2$

$\theta = 2.08$ radians (from GDC)

5 a By cosine rule in triangle PAQ:

$(PQ)^2 = (AP)^2 + (AQ)^2 - 2(AP)(AQ)\cos P\hat{A}Q$

$(PQ)^2 = 6^2 + 6^2 - 2 \times 6 \times 6 \times \dfrac{1}{\sqrt{2}}$

$= 36(2 - \sqrt{2})$

$PQ = 6\sqrt{2 - \sqrt{2}}$

b By cosine rule in triangle PBQ:

$\cos P\hat{B}Q = \dfrac{(PB)^2 + (QB)^2 - (PQ)^2}{2(PB)(QB)}$

$= \dfrac{4^2 + 4^2 - 36(2 - \sqrt{2})}{2(4)(4)}$

$= 0.341$

$P\hat{B}Q = \cos^{-1} 0.341$

$= 70.1° = 1.22$ radians

c Shaded area is the sum of two segments, one of radius 6 and angle $45° = \dfrac{\pi}{4}$ radians and the other of radius 4 and angle 1.22 radians.

Shaded area $= \left[\dfrac{1}{2} \times 6^2 \times \left(\dfrac{\pi}{4} - \sin\left(\dfrac{\pi}{4}\right)\right)\right]$

$+ \left[\dfrac{1}{2} \times 4^2 \times (1.22 - \sin(1.22))\right]$

$= 1.41 + 2.26$

$= 3.67$ cm^2

Mixed examination practice 10

Short questions

1 a

$\hat{COQ} + \dfrac{\pi}{2} + \dfrac{\pi}{6} = \pi$ (angles on a straight line)

$\Rightarrow \hat{COQ} = \dfrac{\pi}{3}$

b

Total area = area COQ + area OABC + area OAP

$= \dfrac{1}{2}\left(2^2 \times \dfrac{\pi}{3}\right) + (2 \times 7) + \dfrac{1}{2}\left(7^2 \times \dfrac{\pi}{6}\right)$

$= \dfrac{57\pi}{12} + 14$

$= 28.9 \text{ cm}^2 \text{ (3SF)}$

c

Perimeter = QC + CB + BA + AP + PO + OQ

$= 2 \times \dfrac{\pi}{3} + 7 + 2 + 7 \times \dfrac{\pi}{6} + 7 + 2$

$= \dfrac{11\pi}{6} + 18 = 23.8 \text{ cm (3SF)}$

2 $p = 2r + r\theta$

$36 = 2 \times 10 + 10\theta$

$\Rightarrow \theta = 1.6$

$\therefore \text{Area} = \dfrac{r^2\theta}{2} = \dfrac{10^2 \times 1.6}{2} = 80 \text{ cm}^2$

3

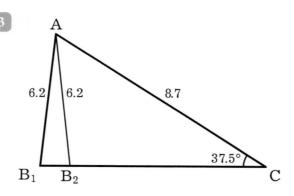

Figure 10MS.3

Using the sine rule,

$\dfrac{\sin A\hat{B}C}{AC} = \dfrac{\sin A\hat{C}B}{AB}$

$\Rightarrow \sin A\hat{B}C = \dfrac{AC \times \sin A\hat{C}B}{AB}$

$= \dfrac{8.7 \sin 37.5°}{6.2}$

$= 0.854$

$A\hat{B}C = \sin^{-1} 0.854$

$= 58.7°$ or $180° - 58.7° = 121°$ (3SF)

4 a $\dfrac{12}{AB} = \tan 56°$

$\therefore AB = \dfrac{12}{\tan 56°} = 8.09 \text{ m (3SF)}$

b Triangle ABM is isosceles, with BM = AB.

Using the cosine rule:

$AM = \sqrt{AB^2 + BM^2 - 2(AB)(BM)\cos A\hat{B}M}$

$= \sqrt{8.09^2 + 8.09^2 - 2 \times 8.09^2 \times \cos 48°}$

$= 6.58 \text{ m}$

5 a Segment area $= \dfrac{1}{2}r^2(\theta - \sin\theta)$

$= \dfrac{1}{2} \times 7^2 (1.4 - \sin 1.4)$

$= 10.2 \text{ cm}^2 \text{ (3SF)}$

b By the cosine rule in triangle OPQ,

$PQ = \sqrt{OP^2 + OQ^2 - 2(OP)(OQ)\cos\theta}$

$= \sqrt{7^2 + 7^2 - 2 \times 7^2 \times \cos 1.4}$

$= 9.02$

Perimeter = length of line PQ
            + length of arc PQ

$= 9.02 + 1.4 \times 7$

$= 18.8 \text{ cm}$

6 a Shaded area is the difference between two sectors:

$$\text{Area} = \frac{10^2 \theta}{2} - \frac{(10-x)^2 \theta}{2}$$

$$= \frac{\theta}{2}(100 - 100 + 20x - x^2)$$

$$= \frac{\theta x(20-x)}{2}$$

b $\theta = 1.2$:

Area $= 54.6$

$$\frac{1.2x(20-x)}{2} = 54.6$$

$0.6x(20-x) = 54.6$

$x^2 - 20x + 91 = 0$

$(x-13)(x-7) = 0$

$\therefore x = 7$ (since $x < 10$)

7 $O\hat{T}A = 90°$ because AT is a tangent.

So, by Pythagoras' Theorem,

$$AT = \sqrt{12^2 - 6^2}$$

$$= \sqrt{108} = 6\sqrt{3} \text{ cm}$$

$\therefore$ Area of triangle OTA $= \frac{1}{2} \times 6\sqrt{3} \times 6$

$$= 18\sqrt{3} \text{ cm}^2$$

$\cos A\hat{O}T = \frac{6}{12}$

$\Rightarrow A\hat{O}T = \frac{\pi}{3}$

Shaded area = area of OTA − area of sector

$$= 18\sqrt{3} - \frac{1}{2} \times 6^2 \times \frac{\pi}{3}$$

$$= 18\sqrt{3} - 6\pi$$

$$= 12.3 \text{ cm}^2$$

8 a

Area of segment BDCP $= \frac{1}{2} \times 2^2 \left(\frac{\pi}{2} - \sin \frac{\pi}{2} \right)$

$$= \pi - 2 = 1.14 \text{ cm}^2$$

b The semicircle with diameter BC has radius $\sqrt{2}$, so area of the region BECD is

area of semicircle − area of segment BDCP

$$= \frac{\pi(\sqrt{2})^2}{2} - (\pi - 2)$$

$$= \pi - (\pi - 2)$$

$$= 2 \text{ cm}^2$$

9 a Angle of sector 2 is $\frac{\pi}{2} - \theta$

$\therefore$ Area of sector 2 $= \frac{2^2}{2} \left(\frac{\pi}{2} - \theta \right)$

$$= \pi - 2\theta$$

b Total removed area is a semicircle with radius 2,

$\therefore$ remaining area $= \frac{9 \times 12}{2} - \frac{\pi \times 2^2}{2}$

$$= 54 - 2\pi = 47.7 \text{ cm}^2$$

10 $p = 34$

$\therefore 2r + r\theta = 34$ ...(1)

Area $= 52$

$\therefore \frac{1}{2} r^2 \theta = 52$ ...(2)

$(2) \Rightarrow \theta = \frac{104}{r^2}$

Substituting into (1):

$2r + \frac{104}{r} = 34$

$r^2 - 17r + 52 = 0$

$(r-13)(r-4) = 0$

$r = 13$ cm or 4 cm

11

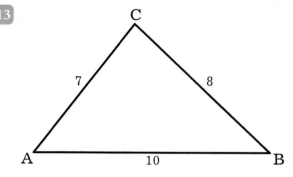

Figure 10MS.11

By the cosine rule,

$$BC^2 = AB^2 + AC^2 - 2(AB)(AC)\cos B\hat{A}C$$

$$BC = \sqrt{(2\sqrt{3})^2 + 10^2 - 2(10)(2\sqrt{3})\cos 150°}$$

$$= \sqrt{12 + 100 - 40\sqrt{3}\left(-\frac{\sqrt{3}}{2}\right)}$$

$$= \sqrt{112 + 60}$$

$$= \sqrt{172}$$

$$= 2\sqrt{43}$$

12 By the sine rule,

$$\frac{\sin L\hat{K}M}{6.1} = \frac{\sin 42°}{4.2}$$

$$\Rightarrow L\hat{K}M = \sin^{-1}\left(\frac{6.1 \sin 42°}{4.2}\right)$$

$$= 76.37°$$

or $180° - 76.37° = 103.63°$

$$\therefore L\hat{M}K = 180° - 42° - L\hat{K}M$$

$$= 61.63° \text{ or } 34.37°$$

Since the triangle is obtuse,

$L\hat{K}M = 103.63°$ and $L\hat{M}K = 34.37°$

$$\text{Area} = \frac{1}{2}(LM)(KM)\sin L\hat{M}K$$

$$= \frac{1}{2}(6.1)(4.2)\sin 34.37°$$

$$= 7.23 \text{ cm}^2 \text{ (3SF)}$$

13

Figure 10MS.13

a By the cosine rule,

$$\cos A\hat{B}C = \frac{AB^2 + BC^2 - AC^2}{2(AB)(BC)}$$

$$= \frac{8^2 + 10^2 - 7^2}{2(8)(10)}$$

$$= \frac{115}{160}$$

$$= \frac{23}{32}$$

b $\sin A\hat{B}C = \sqrt{1 - \cos^2 A\hat{B}C}$

$$= \sqrt{1 - \left(\frac{23}{32}\right)^2}$$

$$= \frac{1}{32}\sqrt{1024 - 529}$$

$$= \frac{1}{32}\sqrt{495}$$

$$= \frac{3}{32}\sqrt{55}$$

c $\text{Area} = \frac{1}{2}(AB)(BC)\sin A\hat{B}C$

$$= \frac{1}{2} \times 10 \times 8 \times \frac{3}{32}\sqrt{55}$$

$$= \frac{15}{4}\sqrt{55} \text{ cm}^2$$

Long questions

1 a

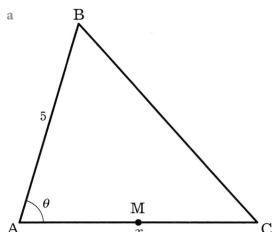

Figure 10ML.1

Using cosine rule in triangle AMB:

$$MB^2 = AM^2 + AB^2 - 2(AM)(AB)\cos M\hat{A}B$$

$$= \left(\frac{x}{2}\right)^2 + 5^2 - 2\left(\frac{x}{2}\right) \times 5\cos\theta$$

$$= \frac{x^2}{4} + 25 - 5x\cos\theta$$

b Using cosine rule in triangle ABC:

$$BC^2 = AC^2 + AB^2 - 2(AB)(BC)\cos B\hat{A}C$$

$$= x^2 + 25 - 10x\cos\theta$$

$BC = MB \Rightarrow BC^2 = MB^2$

i.e. $x^2 + 25 - 10x\cos\theta = \frac{x^2}{4} + 25 - 5x\cos\theta$

$$\frac{3x^2}{4} = 5x\cos\theta$$

$$\therefore \cos\theta = \frac{3x}{20} \quad (\text{as } x \neq 0)$$

c $x = 5 \Rightarrow \cos\theta = \frac{15}{20} = \frac{3}{4}$

$$\therefore \theta = \cos^{-1}\left(\frac{3}{4}\right)$$

$$= 41.4° \text{ (3SF)}$$

2 a

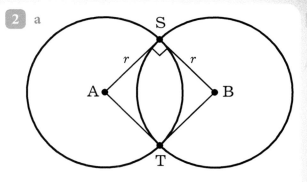

Figure 10ML.2

ASBT is a rhombus as each side has the same length r (the radius).

Since $A\hat{S}B = 90°$, ASBT is a square, and so $S\hat{A}T = 90°$.

b AB is the diagonal of a square with side r.

So, by Pythagoras' Theorem,

$$AB^2 = 2r^2$$
$$AB = \sqrt{2}r$$

c Sector AST is a quarter circle.

$$\therefore \text{Area AST} = \frac{\pi r^2}{4}$$

d The overlap consists of two sectors minus the square ASBT.

$$\text{Area} = 2 \times \frac{\pi r^2}{4} - r^2$$

$$= \frac{r^2}{2}(\pi - 2)$$

3 a Area of minor segment

= area of minor sector AOB
− area of triangle AOB

$$= \frac{r^2\theta}{2} - \frac{1}{2}r^2\sin\theta$$

$$= \frac{r^2}{2}(\theta - \sin\theta)$$

b Area of major sector

= area of circle − area of minor sector

$$= \pi r^2 - \frac{r^2\theta}{2}$$

$$= \frac{r^2}{2}(2\pi - \theta)$$

c $\dfrac{\frac{r^2}{2}(\theta - \sin\theta)}{\frac{r^2}{2}(2\pi - \theta)} = \dfrac{1}{2}$

$2(\theta - \sin\theta) = (2\pi - \theta)$

$2\theta - 2\sin\theta = 2\pi - \theta$

$\sin\theta = \dfrac{3\theta}{2} - \pi$

d From GDC, $\theta = 2.50$ (3SF)

4 a Area $= \dfrac{1}{2}ab\sin C$

$2.21 = \dfrac{1}{2} \times x \times 3x \times \sin\theta$

$\sin\theta = \dfrac{4.42}{3x^2}$

b $(x+3)^2 = x^2 + (3x)^2 - 2(x)(3x)\cos\theta$

$x^2 + 6x + 9 = x^2 + 9x^2 - 6x^2\cos\theta$

$\Rightarrow \cos\theta = \dfrac{9x^2 - 6x - 9}{6x^2}$

$= \dfrac{3x^2 - 2x - 3}{2x^2}$

c $\cos^2\theta = 1 - \sin^2\theta$

$\Rightarrow \left(\dfrac{3x^2 - 2x - 3}{2x^2}\right)^2 = 1 - \left(\dfrac{4.42}{3x^2}\right)^2$

d i Using GDC: $x = 1.24, 2.94$ (3SF)

ii $\theta = \cos^{-1}\left(\dfrac{3x^2 - 2x - 3}{2x^2}\right)$

$= 1.86, 0.172$

5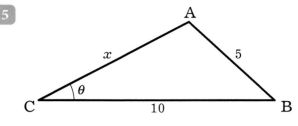

Figure 10ML.5

a By the cosine rule:

$AB^2 = CA^2 + BC^2 - 2(CA)(BC)\cos A\hat{C}B$

$5^2 = x^2 + 10^2 - 2x \times 10\cos\theta = 0$

$25 - x^2 - 100 + 20x\cos\theta = 0$

$x^2 - 20x\cos\theta + 75 = 0$

b Real solutions $\Rightarrow$ discriminant $\Delta \geq 0$:

$(-20\cos\theta)^2 - 4 \times 1 \times 75 \geq 0$

$(20\cos\theta)^2 \geq 300$

$20\cos\theta \leq -10\sqrt{3}$ or $20\cos\theta \geq 10\sqrt{3}$

$\therefore -1 \leq \cos\theta \leq -\dfrac{\sqrt{3}}{2}$ or $\dfrac{\sqrt{3}}{2} \leq \cos\theta \leq 1$

COMMENT

Remember that $-1 \leq \cos\theta \leq 1$ always, so in the absence of other bounds these will still hold.

c $\cos^{-1}(-1) = 180°$

$\cos^{-1}\left(-\dfrac{\sqrt{3}}{2}\right) = 150°$

$\therefore -1 \leq \cos\theta \leq -\dfrac{\sqrt{3}}{2} \Rightarrow 150° \leq \theta < 180°$

$\cos^{-1}\left(\dfrac{\sqrt{3}}{2}\right) = 30°$

$\cos^{-1}(1) = 0°$

$\therefore \dfrac{\sqrt{3}}{2} \leq \cos\theta \leq 1 \Rightarrow 0° < \theta \leq 30°$

i.e. $0° < \theta \leq 30°$ or $150° \leq \theta < 180°$

6 a i $AP^2 = (8-x)^2 + (6-10)^2$
$= 80 - 16x + x^2$
$\Rightarrow AP = \sqrt{x^2 - 16x + 80}$

ii $OP = \sqrt{x^2 + 100}$

b By the cosine rule:

$\cos O\hat{P}A = \dfrac{OP^2 + AP^2 - OA^2}{2(OP)(AP)}$

$= \dfrac{(x^2+100)+(x^2-16x+80)-(8^2+6^2)}{2\sqrt{(x^2+100)(x^2-16x+80)}}$

$= \dfrac{2x^2 - 16x + 80}{2\sqrt{(x^2+100)(x^2-16x+80)}}$

$= \dfrac{x^2 - 8x + 40}{\sqrt{(x^2+100)(x^2-16x+80)}}$

c $x = 8$

$\Rightarrow O\hat{P}A = \cos^{-1}\left(\dfrac{40}{\sqrt{164 \times 16}}\right) = 38.7°$ (3SF)

d $O\hat{P}A = 60° \Rightarrow \cos O\hat{P}A = \dfrac{1}{2}$

$\therefore \dfrac{x^2 - 8x + 40}{\sqrt{(x^2+100)(x^2-16x+80)}} = \dfrac{1}{2}$

From GDC, $x = 5.63$ (3SF)

e i If $f(x) = \cos O\hat{P}A = 1$ then $O\hat{P}A = 0$, which happens when OAP is a straight line (so there is a solution).

ii When OAP forms a straight line, the gradients of OA and OP are equal:

$\dfrac{6-0}{8-0} = \dfrac{10-0}{x-0}$

$\dfrac{6}{8} = \dfrac{10}{x}$

$x = \dfrac{80}{6}$

$= \dfrac{40}{3}$

7 a i $y = -16x^2 + 160x - 256$
$= -16(x^2 - 10x + 16)$
$= -16((x-5)^2 - 25 + 16)$
$= 144 - 16(x-5)^2$

Maximum value of y occurs at $x = 5$

ii Maximum value of y is 144

b i $x + z + 6 = 16$
$\Rightarrow z = 10 - x$

ii $z^2 = x^2 + 6^2 - 2 \times x \times 6 \cos Z$
$= x^2 + 36 - 12x \cos Z$

iii $\cos Z = \dfrac{x^2 + 36 - z^2}{12x}$ from (ii)

$= \dfrac{x^2 + 36 - (10-x)^2}{12x}$ from (i)

$= \dfrac{-64 + 20x}{12x}$

$= \dfrac{5x - 16}{3x}$

c $A = \dfrac{1}{2} \times 6 \times x \times \sin Z$
$= 3x \sin Z$
$\Rightarrow A^2 = 9x^2 \sin^2 Z$

d $A^2 = 9x^2 \sin^2 Z$
$= 9x^2(1 - \cos^2 Z)$
$= 9x^2\left(1 - \left(\dfrac{5x-16}{3x}\right)^2\right)$ by (b)(iii)
$= 9x^2 - 25x^2 + 160x - 256$
$= -16x^2 + 160x - 256$

e i From (a)(ii), the maximum value of A^2 is 144, so the maximum area is 12.

ii From (a)(i), the maximum occurs when $x = 5$, for which $z = 10 - 5 = 5$. Since $x = z$, the triangle is isosceles.

8 a $O_1\hat{A}B = \dfrac{\pi}{2}$, since AB is tangent to the circle.

b By the same reasoning, $O_2\hat{B}A = \dfrac{\pi}{2}$, and hence $BAPO_2$ is a rectangle, so $PO_2 = AB$ (parallel sides in a rectangle have the same length).

c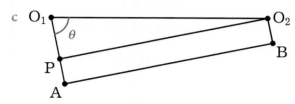

Figure 10ML.8

$PO_1 = 8 - 3 = 5$

$O_1O_2 = 25$

$\therefore PO_2 = \sqrt{25^2 - 5^2}$

$= \sqrt{600}$

$= 10\sqrt{6}$

$AB = PO_2 = 10\sqrt{6} = 24.5$ cm (3SF)

d $\sin\theta = \dfrac{PO_2}{25}$

$= \dfrac{10\sqrt{6}}{25}$

$\theta = \sin^{-1}\left(\dfrac{10\sqrt{6}}{25}\right) = 1.369$ (4SF)

e Length of chain $=$ arc $AD + 2AB +$ arc BC

$= 8(2\pi - 2\theta) + 2 \times 10\sqrt{6} + 3(2\theta)$

$= 85.6$ cm (3SF)

11 Vectors

Exercise 11A

4 a $\overrightarrow{AB} = \boldsymbol{b} - \boldsymbol{a}$

b

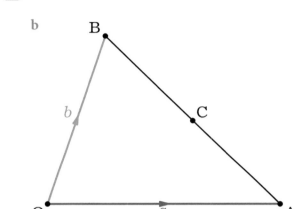

Figure 11A.4

$\overrightarrow{OC} = \overrightarrow{OA} + \dfrac{1}{2}\overrightarrow{AB}$

$= \boldsymbol{a} + \dfrac{1}{2}(\boldsymbol{b} - \boldsymbol{a})$

$= \dfrac{1}{2}(\boldsymbol{a} + \boldsymbol{b})$

c $\overrightarrow{AD} = -3\overrightarrow{AB}$

$\Rightarrow \overrightarrow{OD} = \overrightarrow{OA} + \overrightarrow{AD}$

$= \overrightarrow{OA} - 3\overrightarrow{AB}$

$= \boldsymbol{a} - 3(\boldsymbol{b} - \boldsymbol{a})$

$= 4\boldsymbol{a} - 3\boldsymbol{b}$

5 a $\overrightarrow{AB} = \begin{pmatrix} 4 \\ 2 \end{pmatrix} - \begin{pmatrix} 3 \\ 0 \end{pmatrix} = \begin{pmatrix} 1 \\ 2 \end{pmatrix}$

$\overrightarrow{AC} = \dfrac{1}{2}\overrightarrow{AB} = \begin{pmatrix} 0.5 \\ 1 \end{pmatrix}$

b $\overrightarrow{OD} = \overrightarrow{OA} + \overrightarrow{AD}$

$= \begin{pmatrix} 3 \\ 0 \end{pmatrix} + \begin{pmatrix} 7 \\ -2 \end{pmatrix}$

$= \begin{pmatrix} 10 \\ -2 \end{pmatrix}$

∴ the coordinates of D are (10, −2)

6 a $\overrightarrow{AB} = \overrightarrow{OB} - \overrightarrow{OA}$

$= \begin{pmatrix} 4 \\ -2 \\ 5 \end{pmatrix} - \begin{pmatrix} 3 \\ 1 \\ -2 \end{pmatrix}$

$= \begin{pmatrix} 1 \\ -3 \\ 7 \end{pmatrix}$

b The position vector of the midpoint is the mean of the position vectors of the end points:

$\dfrac{1}{2}\left(\begin{pmatrix} 3 \\ 1 \\ -2 \end{pmatrix} + \begin{pmatrix} 4 \\ -2 \\ 5 \end{pmatrix}\right) = \begin{pmatrix} 3.5 \\ -0.5 \\ 1.5 \end{pmatrix}$

7 $\overrightarrow{OD} = \overrightarrow{OA} + \overrightarrow{AD}$

$= (2\boldsymbol{i} - 3\boldsymbol{j}) + (\boldsymbol{i} - \boldsymbol{j})$

$= 3\boldsymbol{i} - 4\boldsymbol{j}$

8

Figure 11A.8

$\overrightarrow{AC} = \frac{2}{5}\overrightarrow{AB}$

$\overrightarrow{AB} = b - a = \begin{pmatrix} 1 \\ -1 \\ 3 \end{pmatrix} - \begin{pmatrix} 2 \\ 2 \\ 1 \end{pmatrix} = \begin{pmatrix} -1 \\ -3 \\ 2 \end{pmatrix}$

$\therefore \overrightarrow{OC} = \overrightarrow{OA} + \overrightarrow{AC}$

$= \begin{pmatrix} 2 \\ 2 \\ 1 \end{pmatrix} + \frac{2}{5}\begin{pmatrix} -1 \\ -3 \\ 2 \end{pmatrix}$

$= \begin{pmatrix} 1.6 \\ 0.8 \\ 1.8 \end{pmatrix}$

9 a M has position vector

$\frac{1}{2}(p+q) = \frac{1}{2}(2i - j - 3k + i + 4j - k)$

$= \frac{3}{2}i + \frac{3}{2}j - 2k$

b

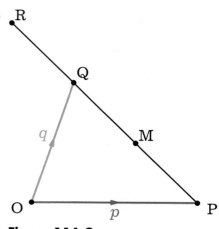

Figure 11A.9

$\overrightarrow{QR} = -\overrightarrow{QM}$

$\therefore \overrightarrow{OR} = \overrightarrow{OQ} + \overrightarrow{QR}$

$= \overrightarrow{OQ} - \overrightarrow{QM}$

$= q - \frac{1}{2}(p - q)$

$= \frac{1}{2}(3q - p)$

Hence $\overrightarrow{OR} = \frac{1}{2}(3(i + 4j - k) - (2i - j - 3k))$

$= \frac{1}{2}i + \frac{13}{2}j$

$\therefore$ the coordinates of R are (0.5, 6.5, 0)

10

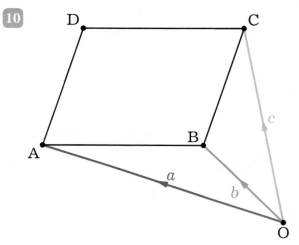

Figure 11A.10

$\overrightarrow{BA} = a - b = \begin{pmatrix} -3 \\ -2 \\ 2 \end{pmatrix}$

To form a parallelogram, $\overrightarrow{CD} = \overrightarrow{BA}$

$\therefore \overrightarrow{OD} = \overrightarrow{OC} + \overrightarrow{CD}$

$= \begin{pmatrix} 3 \\ 1 \\ 4 \end{pmatrix} + \begin{pmatrix} -3 \\ -2 \\ 2 \end{pmatrix}$

$= \begin{pmatrix} 0 \\ -1 \\ 6 \end{pmatrix}$

Exercise 11B

4 $3a + 4x = b$

$\Rightarrow x = \frac{1}{4}(b - 3a)$

$= \frac{1}{4}\left(\begin{pmatrix} 5 \\ 3 \\ 3 \end{pmatrix} - \begin{pmatrix} -3 \\ 3 \\ 6 \end{pmatrix}\right)$

$= \begin{pmatrix} 2 \\ 0 \\ -0.75 \end{pmatrix}$

5 $a + tb = c$

$\Rightarrow tb = c - a$

$t\begin{pmatrix} 1 \\ -1 \\ 2 \end{pmatrix} = \begin{pmatrix} 1 \\ 0 \\ 1 \end{pmatrix} - \begin{pmatrix} 3 \\ -2 \\ 5 \end{pmatrix}$

$t\begin{pmatrix} 1 \\ -1 \\ 2 \end{pmatrix} = \begin{pmatrix} -2 \\ 2 \\ -4 \end{pmatrix}$

$\Rightarrow t = -2$

6 Require that $a + pb = k\begin{pmatrix} 3 \\ 2 \\ 3 \end{pmatrix}$ for some k

i.e. $\begin{pmatrix} 2 + 3p \\ p \\ 2 + 3p \end{pmatrix} = \begin{pmatrix} 3k \\ 2k \\ 3k \end{pmatrix}$

$\therefore p = 2k$ from the second component.

Substituting into $2 + 3p = 3k$:

$2 + 6k = 3k$

$\Rightarrow k = -\frac{2}{3}$

$\therefore p = -\frac{4}{3}$

7 Require that $\lambda x + y = \begin{pmatrix} 0 \\ k \\ 0 \end{pmatrix}$ for some k

i.e. $\begin{pmatrix} 2\lambda + 4 \\ 3\lambda + 1 \\ \lambda + 2 \end{pmatrix} = \begin{pmatrix} 0 \\ k \\ 0 \end{pmatrix}$

$2\lambda + 4 = 0$ from the first component

$\therefore \lambda = -2$

8 Require that $pa + b = k\begin{pmatrix} 1 \\ 1 \\ 2 \end{pmatrix}$ for some k

i.e. $\begin{pmatrix} p + 2q \\ -p + 1 \\ 3p + q \end{pmatrix} = \begin{pmatrix} k \\ k \\ 2k \end{pmatrix}$

$\therefore p + 2q = k$ ...(1)

$-p + 1 = k$ ...(2)

$3p + q = 2k$ ...(3)

$(1) - (2)$:

$2p + 2q - 1 = 0$ ...(4)

$(3) - 2 \times (2)$:

$5p + q - 2 = 0$ ...(5)

Then $2 \times (5) - (4)$:

$8p - 3 = 0$

$\Rightarrow p = \frac{3}{8}$

and hence, substituting into (4):

$q = \frac{1 - 2p}{2} = \frac{1}{8}$

Exercise 11C

7 $\left|\begin{pmatrix} 2c \\ c \\ -c \end{pmatrix}\right| = 12$

$\sqrt{4c^2 + c^2 + c^2} = 12$

$\sqrt{6c^2} = 12$

$6c^2 = 144$

$c^2 = 24$

$c = \pm 2\sqrt{6}$

8 $\vec{AB} = \mathbf{b} - \mathbf{a} = \begin{pmatrix} -2 \\ -2 \\ 1 \end{pmatrix}$

$\Rightarrow AB = \sqrt{2^2 + 2^2 + 1^2} = 3$

$\therefore AC = \dfrac{3}{2}$

9 $\mathbf{a} + \lambda\mathbf{b} = \begin{pmatrix} 2\lambda - 2 \\ -\lambda \\ 2\lambda - 1 \end{pmatrix}$

$|\mathbf{a} + \lambda\mathbf{b}| = 5\sqrt{2}$

$\sqrt{(2\lambda - 2)^2 + \lambda^2 + (2\lambda - 1)^2} = 5\sqrt{2}$

$9\lambda^2 - 12\lambda + 5 = 50$

$3\lambda^2 - 4\lambda - 15 = 0$

$(3\lambda + 5)(\lambda - 3) = 0$

$\lambda = -\dfrac{5}{3}$ or 3

10 a Require $\left|k\begin{pmatrix} 4 \\ -1 \\ 1 \end{pmatrix}\right| = 6$ for some k

$\sqrt{(4^2 + 1^2 + 1^2)k^2} = 6$

$18k^2 = 36$

$\Rightarrow k = \pm\sqrt{2}$

Possible vectors are $\pm\sqrt{2}\begin{pmatrix} 4 \\ -1 \\ 1 \end{pmatrix}$

b Require $\left|k\begin{pmatrix} 2 \\ -1 \\ 1 \end{pmatrix}\right| = 3$ for some $k > 0$

(same direction)

$\sqrt{(2^2 + 1^2 + 1^2)k^2} = 3$

$6k^2 = 9$

$\Rightarrow k = \sqrt{\dfrac{3}{2}} = \dfrac{\sqrt{6}}{2}$ (choose positive root)

$\therefore$ the vector is $\dfrac{\sqrt{6}}{2}\begin{pmatrix} 2 \\ -1 \\ 1 \end{pmatrix}$

11 $\vec{AB} = \vec{OB} - \vec{OA}$

$= \begin{pmatrix} 2 + 2t \\ 4 + t \\ -9 - 5t \end{pmatrix}$

Require that $AB = 3$

$\therefore (2 + 2t)^2 + (4 + t)^2 + (9 + 5t)^2 = 9$

$30t^2 + 106t + 92 = 0$

$15t^2 + 53t + 46 = 0$

$(t + 2)(15t + 23) = 0$

$t = -2$ or $-\dfrac{23}{15} = -1.53 \,(3\text{SF})$

12 $\vec{PQ} = \mathbf{q} - \mathbf{p}$

$= \begin{pmatrix} 2 + t \\ 1 - t \\ 1 + t \end{pmatrix} - \begin{pmatrix} 1 \\ 1 \\ 3 \end{pmatrix} = \begin{pmatrix} t + 1 \\ -t \\ t - 2 \end{pmatrix}$

$(PQ)^2 = (t + 1)^2 + t^2 + (t - 2)^2$

$= 3t^2 - 2t + 5$

$= 3\left(t - \dfrac{1}{3}\right)^2 + \dfrac{14}{3}$

$\Rightarrow$ minimum $(PQ)^2$ is $\dfrac{14}{3}$, at $t = \dfrac{1}{3}$

$\therefore$ minimum PQ is $\sqrt{\dfrac{14}{3}}$, at $t = \dfrac{1}{3}$

Exercise 11D

5 $\vec{AB} = \begin{pmatrix} -3 \\ 5 \\ -1 \end{pmatrix}$, $\vec{OA} = \begin{pmatrix} 2 \\ 2 \\ 3 \end{pmatrix}$

$\theta = \cos^{-1}\left(\dfrac{\vec{AB} \cdot \vec{OA}}{|\vec{AB}||\vec{OA}|}\right)$

$= \cos^{-1}\dfrac{-6+10-3}{\sqrt{9+25+1}\sqrt{4+4+9}}$

$= \cos^{-1}\left(\dfrac{1}{\sqrt{35}\sqrt{17}}\right)$

$= 87.7°$ (3SF)

6 $\vec{AC} = \begin{pmatrix} 4 \\ 0 \\ -1 \end{pmatrix}$, $\vec{BD} = \begin{pmatrix} 6 \\ -4 \\ 1 \end{pmatrix}$

$\theta = \cos^{-1}\left(\dfrac{\vec{AC} \cdot \vec{BD}}{|\vec{AC}||\vec{BD}|}\right)$

$= \cos^{-1}\left(\dfrac{24+0-1}{\sqrt{16+0+1}\sqrt{36+16+1}}\right)$

$= \cos^{-1}\left(\dfrac{23}{\sqrt{17}\sqrt{53}}\right)$

$= 40.0°$ (3SF)

7 a $\vec{AB} = \begin{pmatrix} k \\ 4 \\ 2k \end{pmatrix} - \begin{pmatrix} 2 \\ 4 \\ 1 \end{pmatrix} = \begin{pmatrix} k-2 \\ 0 \\ 2k-1 \end{pmatrix}$

$\vec{BC} = \begin{pmatrix} k+4 \\ 2k+4 \\ 2k+2 \end{pmatrix} - \begin{pmatrix} k \\ 4 \\ 2k \end{pmatrix} = \begin{pmatrix} 4 \\ 2k \\ 2 \end{pmatrix}$

$\vec{DC} = \begin{pmatrix} k+4 \\ 2k+4 \\ 2k+2 \end{pmatrix} - \begin{pmatrix} 6 \\ 2k+4 \\ 3 \end{pmatrix} = \begin{pmatrix} k-2 \\ 0 \\ 2k-1 \end{pmatrix}$

$\vec{AD} = \begin{pmatrix} 6 \\ 2k+4 \\ 3 \end{pmatrix} - \begin{pmatrix} 2 \\ 4 \\ 1 \end{pmatrix} = \begin{pmatrix} 4 \\ 2k \\ 2 \end{pmatrix}$

$\vec{AB} = \vec{DC}$ and $\vec{BC} = \vec{AD}$

$\therefore$ ABCD is a parallelogram.

b When $k = 1$,

$\vec{AB} = \begin{pmatrix} -1 \\ 0 \\ 1 \end{pmatrix}$, $\vec{AD} = \begin{pmatrix} 4 \\ 2 \\ 2 \end{pmatrix}$

$\theta = \cos^{-1}\left(\dfrac{\vec{AB} \cdot \vec{AD}}{|\vec{AB}||\vec{AD}|}\right)$

$= \cos^{-1}\left(\dfrac{-4+0+2}{\sqrt{1+0+1}\sqrt{16+4+4}}\right)$

$= \cos^{-1}\left(\dfrac{-2}{\sqrt{2}\sqrt{24}}\right)$

$= 107°$ (3SF)

The angles of the parallelogram are $107°$ and $180° - \theta = 73.2°$

c For ABCD to be a rectangle, require $\vec{AB} \cdot \vec{AD} = 0$:

$\begin{pmatrix} k-2 \\ 0 \\ 2k-1 \end{pmatrix} \cdot \begin{pmatrix} 4 \\ 2k \\ 2 \end{pmatrix} = 0$

$4k - 8 + 0 + 4k - 2 = 0$

$8k = 10$

$k = \dfrac{5}{4}$

8 $\vec{AB} = \begin{pmatrix} 2 \\ 1 \\ 5 \end{pmatrix}$, $\vec{BC} = \begin{pmatrix} 2 \\ 1 \\ -7 \end{pmatrix}$, $\vec{CA} = \begin{pmatrix} -4 \\ -2 \\ 2 \end{pmatrix}$

a $\vec{AB} \cdot \vec{CA} = -8 - 2 + 10 = 0$

$\Rightarrow B\hat{A}C = 90°$

b $\vec{CB} = \begin{pmatrix} -2 \\ -1 \\ 7 \end{pmatrix}$

$B\hat{C}A = \cos^{-1}\left(\frac{\vec{CB} \cdot \vec{CA}}{|\vec{CB}||\vec{CA}|}\right)$

$= \cos^{-1}\left(\frac{8+2+14}{\sqrt{4+1+49}\sqrt{16+4+4}}\right)$

$= \cos^{-1}\left(\frac{24}{\sqrt{54}\sqrt{24}}\right)$

$= 48.2°$ (3SF)

$\therefore A\hat{B}C = 180° - 90° - 48.2° = 41.8°$

c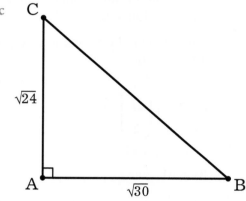

Figure 11D.8

Area $= \frac{1}{2}|AB||CA|$

$= \frac{1}{2}\sqrt{30}\sqrt{24}$

$= 6\sqrt{5}$

$= 13.4$ (3SF)

Exercise 11E

7 a $a \cdot (b+c) = \begin{pmatrix} 2 \\ -2 \\ 1 \end{pmatrix} \cdot \begin{pmatrix} 4 \\ -4 \\ 3 \end{pmatrix} = 19$

b $(b-a) \cdot (d-c) = \begin{pmatrix} -1 \\ 3 \\ 1 \end{pmatrix} \cdot \begin{pmatrix} 0 \\ 2 \\ 1 \end{pmatrix} = 7$

c $(b+d) \cdot (2a) = \begin{pmatrix} 4 \\ -2 \\ 4 \end{pmatrix} \cdot \begin{pmatrix} 4 \\ -4 \\ 2 \end{pmatrix} = 32$

8 a $a \cdot b = 0$ and $a \cdot a = 1$

$\therefore a \cdot (2a - 3b) = 2a \cdot a - 3a \cdot b = 2$

b $\theta = 45° \Rightarrow \cos\theta = \frac{1}{\sqrt{2}}$

$p \cdot q = |p||q|\cos\theta$

$\therefore 3\sqrt{2} = 1 \times |q| \times \frac{1}{\sqrt{2}}$

$\Rightarrow |q| = 6$

9 a $\theta = 60° \Rightarrow \cos\theta = \frac{1}{2}$

$a \cdot b = |a||b|\cos\theta$

$= \frac{3|b|}{2}$

Also, $|a| = 3 \Rightarrow a \cdot a = |a|^2 = 9$

$a \cdot (a - b) = \frac{1}{3}$

$a \cdot a - a \cdot b = \frac{1}{3}$

$9 - \frac{3|b|}{2} = \frac{1}{3}$

$\frac{3|b|}{2} = \frac{26}{3}$

$\Rightarrow |b| = \frac{52}{9}$

b $(3a+b) \cdot (a - 3b) = 0$

$3a \cdot a - 9a \cdot b + b \cdot a - 3b \cdot b = 0$

$3|a|^2 - 3|b|^2 - 8a \cdot b = 0$

Since $|a| = |b|$, the first two terms cancel and this becomes

$-8a \cdot b = 0$

$\therefore a \cdot b = 0$

which means that a and b are perpendicular.

10 a $\overrightarrow{BC} = \begin{pmatrix} -6-2\lambda \\ -\lambda-17 \\ 5 \end{pmatrix}$, $\overrightarrow{AC} = \begin{pmatrix} -7 \\ 4 \\ 2 \end{pmatrix}$,

$\overrightarrow{AB} = \begin{pmatrix} 2\lambda-1 \\ 21+\lambda \\ -3 \end{pmatrix}$

$\overrightarrow{BC} \cdot \overrightarrow{AC} = 0$

$-7(-6-2\lambda) + 4(-\lambda-17) + 10 = 0$

$10\lambda - 16 = 0$

$\lambda = 1.6$

b With $\lambda = 1.6$:

$\overrightarrow{BC} = \begin{pmatrix} -9.2 \\ -18.6 \\ 5 \end{pmatrix}$, $\overrightarrow{AC} = \begin{pmatrix} -7 \\ 4 \\ 2 \end{pmatrix}$, $\overrightarrow{AB} = \begin{pmatrix} 2.2 \\ 22.6 \\ -3 \end{pmatrix}$

$B\hat{C}A = 90°$ from (a)

$B\hat{A}C = \cos^{-1}\left(\dfrac{\overrightarrow{AB} \cdot \overrightarrow{AC}}{|\overrightarrow{AB}||\overrightarrow{AC}|}\right)$

$= \cos^{-1}\left(\dfrac{69}{\sqrt{524.6}\sqrt{69}}\right)$

$= 68.7°$

$B\hat{A}C = 180° - 90° - 68.7° = 21.3°$

c Area $= \dfrac{1}{2}(BC)(AC)$

$= \dfrac{1}{2}\sqrt{455.6}\sqrt{69}$

$= 88.7$ (3SF)

11 a

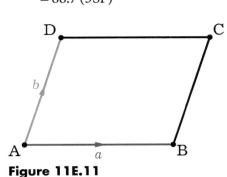

Figure 11E.11

$\overrightarrow{AC} = a + b$

$\overrightarrow{BD} = b - a$

b $(a+b)\cdot(b-a) = a\cdot b - a\cdot a + b\cdot b - a\cdot b$

$= |b|^2 - |a|^2$

c If ABCD is a rhombus, then $|a| = |b|$ and hence $(a+b)\cdot(b-a) = 0$.

This means that $\overrightarrow{AC} \cdot \overrightarrow{BD} = 0$, so the diagonals are perpendicular.

COMMENT

It is very important to know the defining qualities of various quadrilaterals: the square, rhombus, rectangle, parallelogram, trapezium and kite. In a question of this sort, you can assume all other properties without proof, and only need to demonstrate the property being requested.

12 a $\overrightarrow{OB} = \lambda \overrightarrow{OA}$, so $\overrightarrow{OB}$ and $\overrightarrow{OA}$ are parallel, hence B lies on (OA).

b $\overrightarrow{BC} = \begin{pmatrix} 12-2\lambda \\ 2-\lambda \\ 4-4\lambda \end{pmatrix}$, $\overrightarrow{BA} = (1-\lambda)\begin{pmatrix} 2 \\ 1 \\ 4 \end{pmatrix}$

$C\hat{B}A = 90°$

$\Rightarrow \overrightarrow{BC} \cdot \overrightarrow{BA} = 0$

$(1-\lambda)(2(12-2\lambda) + (2-\lambda) + 4(4-4\lambda)) = 0$

$(1-\lambda)(42 - 21\lambda) = 0$

$\therefore \lambda = 2$ (since $\lambda = 1$ is the degenerate case where $\overrightarrow{OA} = \overrightarrow{OB}$)

c

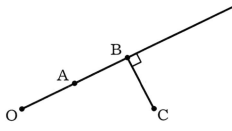

Figure 11E.12

Since $C\hat{B}A = 90°$, B is the point on line (OA) that is closest to C.

Hence the distance from C to the line (OA) is equal to BC:

$$\lambda = 2 \Rightarrow \overrightarrow{BC} = \begin{pmatrix} 8 \\ 0 \\ -4 \end{pmatrix}$$

$$\therefore BC = \left\| \begin{pmatrix} 8 \\ 0 \\ -4 \end{pmatrix} \right\| = \sqrt{64+0+16} = 4\sqrt{5}$$

Exercise 11F

4 Equation of line l: $r = \begin{pmatrix} 2-t \\ 1+t \\ -4+2t \end{pmatrix}$

a At $t = -2$, $r = \begin{pmatrix} 4 \\ -1 \\ -8 \end{pmatrix} = \overrightarrow{OA}$

At $t = 0$, $r = \begin{pmatrix} 2 \\ 1 \\ -4 \end{pmatrix} = \overrightarrow{OB}$

So A and B lie on l.

b At A, $t = -2$, and at B, $t = 0$. Require $AB = BC$, so the point C must lie where $t = 2$.

∴ the coordinates of C are (0, 3, 0)

5 a $\overrightarrow{PQ} = q - p = \begin{pmatrix} -4 \\ -2 \\ 3 \end{pmatrix}$

$\Rightarrow r = \begin{pmatrix} 7 \\ 1 \\ 2 \end{pmatrix} + \lambda \begin{pmatrix} -4 \\ -2 \\ 3 \end{pmatrix}$

b At P, $\lambda = 0$, and at Q, $\lambda = 1$. Require $PR = 2PQ$, so the point R must lie where $\lambda = \pm 2$, since the distances are proportional to the differences in λ.

Hence

$$\overrightarrow{OR} = \begin{pmatrix} 7 \\ 1 \\ 2 \end{pmatrix} + 2\begin{pmatrix} -4 \\ -2 \\ 3 \end{pmatrix} \text{ or } \begin{pmatrix} 7 \\ 1 \\ 2 \end{pmatrix} - 2\begin{pmatrix} -4 \\ -2 \\ 3 \end{pmatrix}$$

i.e. the coordinates of R are (−1, −3, 8) or (15, 5, −4)

6 a $2i - 3j + 6k$ gives a direction vector $\begin{pmatrix} 2 \\ -3 \\ 6 \end{pmatrix}$ for the line,

$$\therefore r = \begin{pmatrix} 2 \\ 1 \\ 4 \end{pmatrix} + \lambda \begin{pmatrix} 2 \\ -3 \\ 6 \end{pmatrix}$$

b $\left\| \begin{pmatrix} 2 \\ -3 \\ 6 \end{pmatrix} \right\| = \sqrt{4+9+36} = 7$

c As P is a point on the line,

$$AP = \left\| \lambda \begin{pmatrix} 2 \\ -3 \\ 6 \end{pmatrix} \right\| = |\lambda| \left\| \begin{pmatrix} 2 \\ -3 \\ 6 \end{pmatrix} \right\| = 7|\lambda|$$

$AP = 35 \Rightarrow |\lambda| = 5 \Rightarrow \lambda = \pm 5$

$$\therefore \overrightarrow{OP} = \begin{pmatrix} 2 \\ 1 \\ 4 \end{pmatrix} + 5\begin{pmatrix} 2 \\ -3 \\ 6 \end{pmatrix} \text{ or } \begin{pmatrix} 2 \\ 1 \\ 4 \end{pmatrix} - 5\begin{pmatrix} 2 \\ -3 \\ 6 \end{pmatrix}$$

i.e. the coordinates of P are (12, −14, 34) or (−8, 16, −26)

Exercise 11G

> **COMMENT**
>
> In these problems, assign unknowns to the values that need to be determined, then use one or more standard equations which describe the geometry you are given. Solving the equations will give the values for the unknowns.

4 Since C lies on l, $\overrightarrow{OC} = \begin{pmatrix} 4+2\lambda \\ 2-\lambda \\ -1+2\lambda \end{pmatrix}$ for some λ

$$\overrightarrow{CP} = p - c = \begin{pmatrix} 3-2\lambda \\ \lambda \\ 4-2\lambda \end{pmatrix}$$

[PC] perpendicular to l

$$\Rightarrow \overrightarrow{CP} \cdot \begin{pmatrix} 2 \\ -1 \\ 2 \end{pmatrix} = 0$$

$6 - 4\lambda - \lambda + 8 - 4\lambda = 0$

$9\lambda = 14$

$\Rightarrow \lambda = \dfrac{14}{9}$

∴ the coordinates of C are $\left(\dfrac{64}{9}, \dfrac{4}{9}, \dfrac{19}{9}\right)$

5 Let P be the point on the line that is closest to $A(-1, 1, 2)$

$$\overrightarrow{OP} = \begin{pmatrix} 1-3t \\ t \\ 2+t \end{pmatrix} \text{ for some } t$$

$$\overrightarrow{PA} = \overrightarrow{OA} - \overrightarrow{OP} = \begin{pmatrix} -2+3t \\ 1-t \\ -t \end{pmatrix} \text{ and we require}$$

that $\overrightarrow{PA} \cdot \begin{pmatrix} -3 \\ 1 \\ 1 \end{pmatrix} = 0$:

$$\begin{pmatrix} -2+3t \\ 1-t \\ -t \end{pmatrix} \cdot \begin{pmatrix} -3 \\ 1 \\ 1 \end{pmatrix} = 0$$

$6 - 9t + 1 - t - t = 0$

$11t = 7$

$t = \dfrac{7}{11}$

$$\therefore PA = \left| \dfrac{1}{11} \begin{pmatrix} -1 \\ 4 \\ -7 \end{pmatrix} \right|$$

$= \dfrac{1}{11}\sqrt{1 + 16 + 49} = \dfrac{\sqrt{66}}{11}$

6 a At P, $\begin{pmatrix} -5-3\lambda \\ 1 \\ 10+4\lambda \end{pmatrix} = \begin{pmatrix} 3+\mu \\ \mu \\ -9+7\mu \end{pmatrix}$, so

$-5 - 3\lambda = 3 + \mu$ …(1)

$1 = \mu$ …(2)

$10 + 4\lambda = -9 + 7\mu$ …(3)

From (2), $\mu = 1$

Substituting into (1) gives

$-5 - 3\lambda = 4 \Rightarrow \lambda = -3$

Then, substituting into (3):
$10 + 4\lambda = -2 = -9 + 7\mu$ is valid, so the lines do intersect.

Substituting, say, $\mu = 1$ into the equation for l_2 gives $(4, 1, -2)$ as the coordinates of P.

> **COMMENT**
>
> Always use all three of the equations to ensure that the solution obtained from two of them is valid. In this case, the question states that the lines intersect, so this check serves to reveal errors in working rather than determining whether or not the lines are skew, but it is nonetheless worthwhile.

b When $\mu = 2$, $r_2 = \begin{pmatrix} 5 \\ 2 \\ 5 \end{pmatrix}$, so $Q(5, 2, 5)$ does lie on l_2.

c $\overrightarrow{QM} = \begin{pmatrix} -5-3\lambda \\ 1 \\ 10+4\lambda \end{pmatrix} - \begin{pmatrix} 5 \\ 2 \\ 5 \end{pmatrix} = \begin{pmatrix} -10-3\lambda \\ -1 \\ 5+4\lambda \end{pmatrix}$ for some value of λ

Require that $\overrightarrow{QM} \cdot \begin{pmatrix} -3 \\ 0 \\ 4 \end{pmatrix} = 0$:

$$\begin{pmatrix} -10-3\lambda \\ -1 \\ 5+4\lambda \end{pmatrix} \cdot \begin{pmatrix} -3 \\ 0 \\ 4 \end{pmatrix} = 0$$

$30 + 9\lambda + 20 + 16\lambda = 0$

$25\lambda = -50$

$\lambda = -2$

∴ the coordinates of M are $(1, 1, 2)$

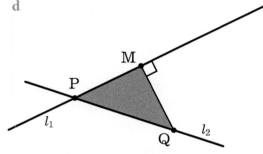

Figure 11G.6

$P\hat{M}Q = 90°$

∴ Area of PMQ $= \dfrac{1}{2}(PM)(MQ)$

$= \dfrac{1}{2}\left\| \begin{pmatrix} -3 \\ 0 \\ 4 \end{pmatrix} \right\| \left\| \begin{pmatrix} -4 \\ -1 \\ -3 \end{pmatrix} \right\|$

$= \dfrac{5\sqrt{26}}{2}$

$= 12.7 \,(3\text{SF})$

7 Let P be the closest point to the origin on the line.

$\overrightarrow{OP} = \begin{pmatrix} 1+2\lambda \\ -2+2\lambda \\ 2+\lambda \end{pmatrix}$ for some value of λ

Require that $\overrightarrow{OP} \cdot \begin{pmatrix} 2 \\ 2 \\ 1 \end{pmatrix} = 0$:

$\begin{pmatrix} 1+2\lambda \\ -2+2\lambda \\ 2+\lambda \end{pmatrix} \cdot \begin{pmatrix} 2 \\ 2 \\ 1 \end{pmatrix} = 0$

$2+4\lambda - 4+4\lambda + 2+\lambda = 0$

$9\lambda = 0$

$\lambda = 0$

∴ $P = (1, -2, 2)$

$OP = \sqrt{1+4+4} = 3$

8 a At P, $\begin{pmatrix} \lambda \\ -1+5\lambda \\ 2+3\lambda \end{pmatrix} = \begin{pmatrix} 2-t \\ 2+t \\ 1+3t \end{pmatrix}$, so

$\lambda = 2-t$ ...(1)

$-1+5\lambda = 2+t$ ...(2)

$2+3\lambda = 1+3t$ ...(3)

(1)+(2):

$6\lambda - 1 = 4 \Rightarrow \lambda = \dfrac{5}{6}$

Substituting into (1):

$t = 2 - \lambda = \dfrac{7}{6}$

Substituting into (3):

$2 + 3\lambda = \dfrac{9}{2} = 1+3t$ is valid, so the lines do intersect.

Substituting, say, $\lambda = \dfrac{5}{6}$ into the equation for l_1 gives $\left(\dfrac{5}{6}, \dfrac{19}{6}, \dfrac{27}{6}\right)$ as the coordinates of P.

b $\boldsymbol{d}_1 = \begin{pmatrix} 1 \\ 5 \\ 3 \end{pmatrix}$, $\boldsymbol{d}_2 = \begin{pmatrix} -1 \\ 1 \\ 3 \end{pmatrix}$

Let θ be the angle between the lines. Then

$\theta = \cos^{-1}\left(\dfrac{\boldsymbol{d}_1 \cdot \boldsymbol{d}_2}{|\boldsymbol{d}_1||\boldsymbol{d}_2|}\right)$

$= \cos^{-1}\left(\dfrac{-1+5+9}{\sqrt{1+25+9}\sqrt{1+1+9}}\right)$

$= \cos^{-1}\left(\dfrac{13}{\sqrt{35}\sqrt{11}}\right)$

$= 48.5°\,(3\text{SF})$

c At $t = 3$, $\boldsymbol{r}_2 = \begin{pmatrix} -1 \\ 5 \\ 10 \end{pmatrix} = \overrightarrow{OQ}$, so Q lies on l_2.

112 Topic 11G Solving problems involving lines

d $\overrightarrow{PQ} = \begin{pmatrix} -1 \\ 5 \\ 10 \end{pmatrix} - \frac{1}{6}\begin{pmatrix} 5 \\ 19 \\ 27 \end{pmatrix} = \frac{1}{6}\begin{pmatrix} -11 \\ 11 \\ 33 \end{pmatrix} = \frac{11}{6}\begin{pmatrix} -1 \\ 1 \\ 3 \end{pmatrix}$

$\therefore PQ = \frac{11}{6}\sqrt{1+1+9}$

$= \frac{11\sqrt{11}}{6}$

e

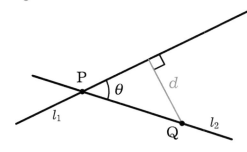

Figure 11G.8

Let R be the closest point to Q on line l_1.

Then PQR is a triangle with

$P\hat{R}Q = 90°$, $PQ = \frac{11\sqrt{11}}{6}$ and $Q\hat{P}R = 48.5°$

By trigonometry in PQR,

$QR = PQ \sin Q\hat{P}R$

$= 4.55 \,(3SF)$

9 a $\overrightarrow{PM} = \begin{pmatrix} 5+2\lambda \\ 1-3\lambda \\ 2+3\lambda \end{pmatrix} - \begin{pmatrix} 21 \\ 5 \\ 10 \end{pmatrix} = \begin{pmatrix} -16+2\lambda \\ -4-3\lambda \\ -8+3\lambda \end{pmatrix}$ for some value of λ

Require $\overrightarrow{PM} \cdot \begin{pmatrix} 2 \\ -3 \\ 3 \end{pmatrix} = 0$

$\therefore \begin{pmatrix} -16+2\lambda \\ -4-3\lambda \\ -8+3\lambda \end{pmatrix} \cdot \begin{pmatrix} 2 \\ -3 \\ 3 \end{pmatrix} = 0$

$-32 + 4\lambda + 12 + 9\lambda - 24 + 9\lambda = 0$

$22\lambda = 44$

$\lambda = 2$

$\therefore$ the coordinates of M are $(9, -5, 8)$

b When $\lambda = 5$, $\boldsymbol{r} = \begin{pmatrix} 15 \\ -14 \\ 17 \end{pmatrix} = \overrightarrow{OQ}$, so Q lies on l.

c

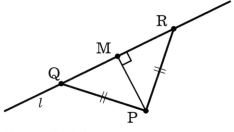

Figure 11G.9

Require PQR to be an isosceles triangle, with base lying on l.

It follows that M must lie at the midpoint of the base of the triangle, since (PM) is perpendicular to (QR).

M has $\lambda = 2$ and Q has $\lambda = 5$, so R must have $\lambda = -1$ for M to be the midpoint of [QR].

$\therefore$ the coordinates of R are $(3, 4, -1)$

10 a At $\mu = 3$, $\boldsymbol{r}_2 = \begin{pmatrix} 5 \\ 2 \\ 6 \end{pmatrix} = \overrightarrow{OQ}$, so Q lies on l_2.

b By inspection, the two lines share a common position vector

$\therefore P = (2, -1, 0)$

> **COMMENT**
>
> Here you could carry out the standard procedure for finding the point of intersection, but it is much easier if you spot the common position vector!

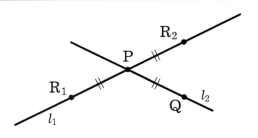

Figure 11G.10

Require PQR to be an isosceles triangle with PR = PQ (from Figure 11G.10 there are two possible positions for R).

$$PQ = \left\|\begin{pmatrix} 3 \\ 3 \\ 6 \end{pmatrix}\right\| = 3\sqrt{6}$$

$$|d_1| = \left\|\begin{pmatrix} 1 \\ -2 \\ 2 \end{pmatrix}\right\| = 3 \Rightarrow PR = 3|\lambda|$$

$$\therefore 3|\lambda| = 3\sqrt{6}$$
$$\Rightarrow \lambda = \pm\sqrt{6}$$

∴ the coordinates of R are

$$\left(2+\sqrt{6},\ -1-2\sqrt{6},\ 2\sqrt{6}\right)$$
or $\left(2-\sqrt{6},\ -1+2\sqrt{6},\ -2\sqrt{6}\right)$

Mixed examination practice 11
Short questions

1 Direction vector is $\begin{pmatrix} 6 \\ 0 \\ 1 \end{pmatrix} - \begin{pmatrix} 3 \\ -1 \\ 1 \end{pmatrix} = \begin{pmatrix} 3 \\ 1 \\ 0 \end{pmatrix}$

∴ equation is $r = \begin{pmatrix} 3 \\ -1 \\ 1 \end{pmatrix} + \lambda \begin{pmatrix} 3 \\ 1 \\ 0 \end{pmatrix}$

2 a $\overrightarrow{MD} = \overrightarrow{MC} + \overrightarrow{CD}$

$$= \frac{1}{2}\overrightarrow{BC} + \overrightarrow{CD}$$

$$= \frac{1}{2}\overrightarrow{AD} - \overrightarrow{AB}$$

b Since ABCD is a rectangle, $\overrightarrow{AB} \cdot \overrightarrow{AD} = 0$

$$\overrightarrow{MD} \cdot \overrightarrow{MC} = \left(\frac{1}{2}\overrightarrow{AD} - \overrightarrow{AB}\right) \cdot \left(\frac{1}{2}\overrightarrow{AD}\right)$$

$$= \frac{1}{4}\overrightarrow{AD}^2 - \frac{1}{2}\overrightarrow{AB} \cdot \overrightarrow{AD}$$

$$= \frac{1}{4} \times 4^2 - \frac{1}{2} \times 0$$

$$= 4$$

3 Point A corresponds to $\lambda = 0$, and point B corresponds to $\lambda = 2$

AP = 3AB ⇒ P corresponds to $\lambda = 6$
∴ the coordinates of P are (11, 13, 8)

4 $r = \begin{pmatrix} -3 \\ 0 \\ 4 \end{pmatrix} + \lambda \begin{pmatrix} 2 \\ 2 \\ -1 \end{pmatrix}$

$$|d| = \sqrt{2^2 + 2^2 + 1} = 3$$

Point A corresponds to $\lambda = 0$

So a point that is 10 units from A has $\lambda = \pm\dfrac{10}{3}$

The possible coordinates are

$$\begin{pmatrix} -3 \\ 0 \\ 4 \end{pmatrix} \pm \frac{10}{3}\begin{pmatrix} 2 \\ 2 \\ -1 \end{pmatrix} = \left(\frac{11}{3},\ \frac{20}{3},\ \frac{2}{3}\right)$$

and $\left(-\dfrac{29}{3},\ -\dfrac{20}{3},\ \dfrac{22}{3}\right)$

5 a $\begin{pmatrix} 8 \\ -11 \\ 20 \end{pmatrix} - \begin{pmatrix} 4 \\ 1 \\ 12 \end{pmatrix} = \begin{pmatrix} 4 \\ -12 \\ 8 \end{pmatrix}$, so a direction

vector for the line is $d = \begin{pmatrix} 1 \\ -3 \\ 2 \end{pmatrix}$

∴ the line has equation

$$r = \begin{pmatrix} 4 \\ 1 \\ 12 \end{pmatrix} + \lambda \begin{pmatrix} 1 \\ -3 \\ 2 \end{pmatrix}$$

b Since point P lies on l, $\overrightarrow{OP} = \begin{pmatrix} 4+\lambda \\ 1-3\lambda \\ 12+2\lambda \end{pmatrix}$

for some value of λ

Require that $\overrightarrow{OP} \cdot d = 0$:

$\begin{pmatrix} 4+\lambda \\ 1-3\lambda \\ 12+2\lambda \end{pmatrix} \cdot \begin{pmatrix} 1 \\ -3 \\ 2 \end{pmatrix} = 0$

$(4-3+24) + \lambda(1+9+4) = 0$

$25 + 14\lambda = 0$

$\lambda = -\dfrac{25}{14}$

∴ the coordinates of P are $\left(\dfrac{31}{14}, \dfrac{89}{14}, \dfrac{118}{14}\right)$

6 Choosing H to be the origin, $\overrightarrow{HG}$ as the positive x direction, $\overrightarrow{EH}$ as the positive y direction and $\overrightarrow{DH}$ as the positive z direction, we have

$\overrightarrow{HA} = \begin{pmatrix} 0 \\ -4 \\ -3 \end{pmatrix}$ and $\overrightarrow{HC} = \begin{pmatrix} 6 \\ 0 \\ -3 \end{pmatrix}$.

$A\hat{H}C = \cos^{-1}\left(\dfrac{\overrightarrow{HA} \cdot \overrightarrow{HC}}{|\overrightarrow{HA}||\overrightarrow{HC}|}\right)$

$= \cos^{-1}\left(\dfrac{9}{\sqrt{25}\sqrt{45}}\right)$

$= 74.4°$ (3SF)

7 $|a| = |b| = \sqrt{\sin^2\theta + \cos^2\theta} = 1$

$\alpha = \cos^{-1}\left(\dfrac{a \cdot b}{|a||b|}\right)$

$= \cos^{-1}\left(\dfrac{\cos\theta\sin\theta + \sin\theta\cos\theta}{1 \times 1}\right)$

$= \cos^{-1}(2\cos\theta\sin\theta)$

$= \cos^{-1}(\sin 2\theta)$

$= \cos^{-1}\left(\cos\left(\dfrac{\pi}{2} - 2\theta\right)\right)$

$= \dfrac{\pi}{2} - 2\theta$

8 $a = \dfrac{1}{2}((a+b)+(a-b))$,

$b = \dfrac{1}{2}((a+b)-(a-b))$

$a \cdot b = \dfrac{1}{4}((a+b)+(a-b)) \cdot ((a+b)-(a-b))$

$= \dfrac{1}{4}[(a+b) \cdot (a+b) - (a+b)(a-b)$

$+ (a-b)(a+b) - (a-b) \cdot (a-b)]$

$= \dfrac{1}{4}\left(|a+b|^2 - |a-b|^2\right)$

$= 0$

9 a $(b-a) \cdot (b-a) = b \cdot b - b \cdot a - a \cdot b + a \cdot a$

$= |b|^2 + |a|^2 - 2a \cdot b$

b

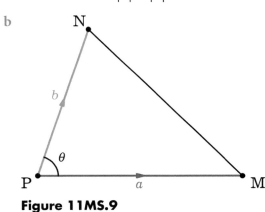

Figure 11MS.9

$b - a = \overrightarrow{MN}$, $|a| = PM$, $|b| = PN$

From (a):

$|b-a|^2 = |a|^2 + |b|^2 - 2a \cdot b$

$= |a|^2 + |b|^2 - 2|a||b|\cos\theta$

i.e. $MN^2 = PM^2 + PN^2 - 2(PM)(PN)\cos\theta$

Long questions

1 a $\overrightarrow{AD} = \begin{pmatrix} 3 \\ 1 \\ k \end{pmatrix} - \begin{pmatrix} 1 \\ 1 \\ 7 \end{pmatrix}$

$= \begin{pmatrix} 2 \\ 0 \\ k-7 \end{pmatrix}$

b (AD) is perpendicular to (AB)
$$\Rightarrow \overrightarrow{AD} \cdot \overrightarrow{AB} = 0$$
$$\begin{pmatrix} 2 \\ 0 \\ k-7 \end{pmatrix} \cdot \begin{pmatrix} -2 \\ 5 \\ -4 \end{pmatrix} = 0$$
$$-4 - 4(k-7) = 0$$
$$4k = 24$$
$$k = 6$$

c With $k = 6$, $\overrightarrow{AD} = \begin{pmatrix} 2 \\ 0 \\ -1 \end{pmatrix}$

$$\therefore \overrightarrow{BC} = \begin{pmatrix} 4 \\ 0 \\ -2 \end{pmatrix}$$

$$c - b = \begin{pmatrix} 4 \\ 0 \\ -2 \end{pmatrix}$$

$$\Rightarrow c = \begin{pmatrix} 4 \\ 0 \\ -2 \end{pmatrix} + \begin{pmatrix} -1 \\ 6 \\ 3 \end{pmatrix} = \begin{pmatrix} 3 \\ 6 \\ 1 \end{pmatrix}$$

i.e. the coordinates of C are (3, 6, 1)

d $\cos(A\hat{D}C) = \dfrac{\overrightarrow{DA} \cdot \overrightarrow{DC}}{DA \times DC}$

$$= \dfrac{\begin{pmatrix} -2 \\ 0 \\ 1 \end{pmatrix} \cdot \begin{pmatrix} 0 \\ 5 \\ -5 \end{pmatrix}}{\sqrt{5} \sqrt{50}}$$

$$= -\dfrac{1}{\sqrt{10}}$$

2 a When $t = 1$, $r_2 = \begin{pmatrix} 4 \\ 1 \\ 2 \end{pmatrix} = \overrightarrow{OA}$, so A lies on l_2.

b $AB = \left\| \begin{pmatrix} -4 \\ 4 \\ -1 \end{pmatrix} \right\| = \sqrt{33}$

c $d_1 = \begin{pmatrix} 2 \\ -1 \\ 3 \end{pmatrix}$, $d_2 = \begin{pmatrix} 4 \\ -4 \\ 1 \end{pmatrix}$

Angle between l_1 and l_2 is

$$\theta = \cos^{-1}\left(\dfrac{d_1 \cdot d_2}{|d_1||d_2|} \right)$$

$$= \cos^{-1}\left(\dfrac{8+4+3}{\sqrt{4+1+9} \sqrt{16+16+1}} \right)$$

$$= \cos^{-1}\left(\dfrac{15}{\sqrt{14} \sqrt{33}} \right)$$

$$= 45.7° \text{ (3SF)}$$

d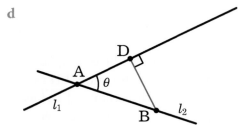

Figure 11ML.2

If D is the point on l_1 that is closest to B, then triangle ABD has
$AB = \sqrt{33}$, $A\hat{D}B = 90°$, $D\hat{A}B = \theta$

$\therefore BD = AB \sin\theta = 4.11 \text{ (3SF)}$

3 a $\overrightarrow{CD} = \overrightarrow{CB} + \overrightarrow{BD}$

$$= \overrightarrow{CB} + \dfrac{k}{k+1} \overrightarrow{BA}$$

$$= \begin{pmatrix} 2 \\ 3 \\ -4 \end{pmatrix} + \dfrac{k}{k+1} \begin{pmatrix} -3 \\ -3 \\ 0 \end{pmatrix}$$

b $\overrightarrow{CD} \cdot \overrightarrow{AB} = 0$

$$\left(\begin{pmatrix} 2 \\ 3 \\ -4 \end{pmatrix} + \dfrac{k}{k+1} \begin{pmatrix} -3 \\ -3 \\ 0 \end{pmatrix} \right) \cdot \begin{pmatrix} 3 \\ 3 \\ 0 \end{pmatrix} = 0$$

$$15 - \dfrac{18k}{k+1} = 0$$

$$15k + 15 - 18k = 0$$

$$k = 5$$

c With $k = 5$,

$$\overrightarrow{CD} = \begin{pmatrix} 2 \\ 3 \\ -4 \end{pmatrix} + \frac{5}{6}\begin{pmatrix} -3 \\ -3 \\ 0 \end{pmatrix} = \begin{pmatrix} -\frac{1}{2} \\ \frac{1}{2} \\ -4 \end{pmatrix}$$

i.e. $\mathbf{d} - \mathbf{c} = \begin{pmatrix} -\frac{1}{2} \\ \frac{1}{2} \\ -4 \end{pmatrix}$

$$\Rightarrow \mathbf{d} = \begin{pmatrix} -\frac{1}{2} \\ \frac{1}{2} \\ -4 \end{pmatrix} + \begin{pmatrix} 2 \\ 1 \\ 6 \end{pmatrix} = \begin{pmatrix} \frac{3}{2} \\ \frac{3}{2} \\ 2 \end{pmatrix}$$

∴ the coordinates of D are $\left(\frac{3}{2}, \frac{3}{2}, 2\right)$

d $CD = \sqrt{\frac{1}{4} + \frac{1}{4} + 16}$

$= \sqrt{\frac{33}{2}}$

$= 4.06$ (3SF)

4 a $P = (a, a^2)$

b $\overrightarrow{PO} = -\overrightarrow{OP} = \begin{pmatrix} -a \\ -a^2 \end{pmatrix}$

$\overrightarrow{PS} = \begin{pmatrix} 0 \\ 4 \end{pmatrix} - \begin{pmatrix} a \\ a^2 \end{pmatrix} = \begin{pmatrix} -a \\ 4-a^2 \end{pmatrix}$

c $\overrightarrow{PO} \cdot \overrightarrow{PS} = 0$

$a^2 - 4a^2 + a^4 = 0$

$a^2(-3 + a^2) = 0$

$a = 0$ or $\pm\sqrt{3}$

Since $a > 0$, $a = \sqrt{3}$

d With $a = \sqrt{3}$, we know that
$\widehat{OPS} = 90°$, $OP = \sqrt{12}$ and $PS = 2$

∴ Area of $OPS = \frac{1}{2} \times \sqrt{12} \times 2 = 2\sqrt{3}$

5 a $\mathbf{r}_1 = \begin{pmatrix} 3t \\ 4t \end{pmatrix}$

b $\mathbf{r}_2 = \begin{pmatrix} 0 \\ 18 \end{pmatrix} + t\begin{pmatrix} 3 \\ -5 \end{pmatrix}$

$= \begin{pmatrix} 3t \\ 18 - 5t \end{pmatrix}$

c $d = |\mathbf{r}_2 - \mathbf{r}_1|$

$= \left|\begin{pmatrix} 0 \\ 18 - 9t \end{pmatrix}\right|$

$= |18 - 9t|$

When $t = 0.5$, $d = 13.5$

d The ships meet if there is a positive value of t for which $d = 0$:

$18 - 9t = 0$

$\Rightarrow t = 2$

∴ the ships do meet, and this happens after 2 hours.

e $d = 18$ km when

$|18 - 9t| = 18$

$9t - 18 = 18$ (since $t > 2$)

$9t = 36$

$t = 4$

∴ the ships are 18 km apart a further 2 hours after meeting.

COMMENT

Note that the other solution to the modulus equation $|18 - 9t| = 18$ is $18 - 9t = 18 \Rightarrow t = 0$, which is already known since the ships start 18 km apart.

6 a $\mathbf{r}_1 = \begin{pmatrix} 0 \\ 5 \\ 0 \end{pmatrix} + t\begin{pmatrix} 3 \\ -4 \\ 1 \end{pmatrix}$

$= \begin{pmatrix} 3t \\ 5 - 4t \\ t \end{pmatrix}$

b $r_2 = \begin{pmatrix} 0 \\ 0 \\ 7 \end{pmatrix} + t \begin{pmatrix} 5 \\ 2 \\ -1 \end{pmatrix}$

$= \begin{pmatrix} 5t \\ 2t \\ 7-t \end{pmatrix}$

$d = |r_1 - r_2|$

$= \sqrt{(2t)^2 + (6t-5)^2 + (7-2t)^2}$

$\Rightarrow d^2 = 4t^2 + 36t^2 - 60t + 25 + 49 - 28t + 4t^2$

$= 44t^2 - 88t + 74$

c $d^2 = 44(t-1)^2 + 30 \geq 30$

i.e. $d \neq 0$ for any t and so the aircraft do not collide.

d From (c), the minimum d^2 is 30, so the minimum distance d is $\sqrt{30} = 5.48$ km (3SF)

7 a Suppose that the two lines intersect; then at the intersection point,

$\begin{pmatrix} -3 \\ 3 \\ 18 \end{pmatrix} + \lambda \begin{pmatrix} 2 \\ -1 \\ -8 \end{pmatrix} = \begin{pmatrix} 5 \\ 0 \\ 2 \end{pmatrix} + \mu \begin{pmatrix} 1 \\ 1 \\ -1 \end{pmatrix}$

$2\lambda - \mu = 8$ ...(1)
$-\lambda - \mu = -3$ ...(2)
$-8\lambda + \mu = -16$ ...(3)

(1)+(3):

$-6\lambda = -8$

$\Rightarrow \lambda = \frac{4}{3}$

(2)+(3):

$-9\lambda = -19$

$\Rightarrow \lambda = \frac{19}{9}$

These two values of λ are not the same, so there is no single value of λ that satisfies all the equations; hence there is no intersection point.

b i $\overrightarrow{PQ} = \begin{pmatrix} 5+\mu \\ \mu \\ 2-\mu \end{pmatrix} - \begin{pmatrix} -3+2\lambda \\ 3-\lambda \\ 18-8\lambda \end{pmatrix}$

$= \begin{pmatrix} \mu - 2\lambda + 8 \\ \mu + \lambda - 3 \\ -\mu + 8\lambda - 16 \end{pmatrix}$

ii (PQ) is perpendicular to l_1

$\Rightarrow \overrightarrow{PQ} \cdot \begin{pmatrix} 2 \\ -1 \\ -8 \end{pmatrix} = 0$

$2(\mu - 2\lambda + 8) - (\mu + \lambda - 3) - 8(-\mu + 8\lambda - 16) = 0$

$9\mu - 69\lambda + 147 = 0$

iii (PQ) is perpendicular to l_2

$\Rightarrow \overrightarrow{PQ} \cdot \begin{pmatrix} 1 \\ 1 \\ -1 \end{pmatrix} = 0$

$(\mu - 2\lambda + 8) + (\mu + \lambda - 3) - (-\mu + 8\lambda - 16) = 0$

$3\mu - 9\lambda + 21 = 0$

iv $9\mu - 69\lambda = -147$ ...(1)

$3\mu - 9\lambda = -21$ ...(2)

$3 \times (2) - (1)$:

$42\lambda = 84$

$\Rightarrow \lambda = 2$

Substituting into (2):

$3\mu - 18 = -21$

$\Rightarrow \mu = -1$

$\therefore P = (1, 1, 2)$ and $Q = (4, -1, 3)$

v The shortest distance between l_1 and l_2 is PQ:

$PQ = \sqrt{3^2 + 2^2 + 1^2}$

$= \sqrt{14}$

$= 3.74$ (3SF)

8 a Parallel lines have the same direction vector

∴ the equation of line L is

$$r = \begin{pmatrix} -2 \\ 4 \\ 2 \end{pmatrix} + \lambda \begin{pmatrix} 1 \\ 1 \\ 0 \end{pmatrix}$$

b $\overrightarrow{AB} = \begin{pmatrix} 4 \\ -1 \\ 1 \end{pmatrix}$

Angle θ between (AB) and L has

$$\cos \theta = \frac{l \cdot \overrightarrow{AB}}{|l| \times AB}$$

$$= \frac{3}{\sqrt{2}\sqrt{18}}$$

$$= \frac{1}{2}$$

c $AB = \sqrt{4^2 + (-1)^2 + 1^2}$

$= \sqrt{18} = 3\sqrt{2}$

$= 4.24$ (3SF)

d

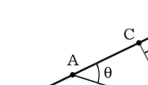

Figure 11ML.8

$AC = AB \cos \theta$

$= 3\sqrt{2} \times \dfrac{1}{2}$

$= \dfrac{3\sqrt{2}}{2}$

12 Basic differentiation and its applications

Exercise 12A

3 a Sometimes true: the derivative indicates the slope of the curve, not its position.
For example, $y = 2x$ has constant positive gradient 2, both at point (1, 2), where $y > 0$, and at point (−1, −2), where $y < 0$.

b Sometimes true: as for (a).
For example, (1, −2) lies on the line $y = -2x$ (with negative gradient) and also on $y = x - 3$ (with positive gradient).

c Always true: $\dfrac{dy}{dx} = 0$ is a defining property of a stationary point.

d Sometimes true: there is also the possibility of a horizontal inflexion point.
For example, $\dfrac{dx^3}{dx}(0) = 0$, but (0, 0) is neither a local maximum nor a local minimum of the curve $y = x^3$.

e Sometimes true; for example, the function $y = -e^{-x}$ has a positive gradient throughout, but its graph is always below the x-axis.

f Sometimes true; for example, the lowest value of the function $y = \sqrt{x}$ is 0, at $x = 0$, but the gradient at $x = 0$ is not zero.

Exercise 12B

2 If $f(x) = x^2 + 1$ then $f(x+h) = (x+h)^2 + 1$.

$$\dfrac{dy}{dx} = \lim_{h \to 0} \left\{ \dfrac{(x+h)^2 + 1 - (x^2 + 1)}{h} \right\}$$

$$= \lim_{h \to 0} \left\{ \dfrac{x^2 + 2xh + h^2 + 1 - x^2 - 1}{h} \right\}$$

$$= \lim_{h \to 0} \left\{ \dfrac{2xh + h^2}{h} \right\}$$

$$= \lim_{h \to 0} \{ 2x + h \}$$

$$= 2x$$

3 If $f(x) = 8$ then $f(x+h) = 8$ as well.

$$\dfrac{dy}{dx} = \lim_{h \to 0} \left\{ \dfrac{8 - 8}{h} \right\}$$

$$= \lim_{h \to 0} \{ 0 \}$$

$$= 0$$

4 Let $y = kf(x)$.

$$\dfrac{dy}{dx} = \lim_{h \to 0} \left\{ \dfrac{kf(x+h) - kf(x)}{h} \right\}$$

$$= \lim_{h \to 0} \left\{ k \dfrac{f(x+h) - f(x)}{h} \right\}$$

$$= k \lim_{h \to 0} \left\{ \dfrac{f(x+h) - f(x)}{h} \right\}$$

$$= k f'(x)$$

Exercise 12D

8 $\frac{dy}{dx} = 3x^2 + k$

Since $3x^2 \geq 0$ for all values of x, if $k > 0$ then $\frac{dy}{dx} > 0$ for all values of x.

That is, $y(x)$ is always increasing if $k > 0$.

9 $\frac{dy}{dx} = 3x^2 - 4x$

$\frac{dy}{dx} = y$

$3x^2 - 4x = x^3 - 2x^2 + 1$

$x^3 - 5x^2 + 4x + 1 = 0$

From GDC: $x = -0.199, 1.29, 3.91$ (3SF)

The points are $(-0.199, 0.913)$, $(1.29, -0.181)$, $(3.91, 30.3)$

10 The gradient is decreasing where $\frac{d^2y}{dx^2} < 0$

$\frac{dy}{dx} = 7 - 2x - 3x^2$

$\therefore \frac{d^2y}{dx^2} = -2 - 6x$

$-2 - 6x < 0$

$\Rightarrow 6x > -2$

i.e. $x > -\frac{1}{3}$

11 $\frac{d^n}{dx^n}(x^n) = \frac{d^{n-1}}{dx^{n-1}}\left(\frac{d}{dx}(x^n)\right)$

$= \frac{d^{n-1}}{dx^{n-1}}(nx^{n-1})$

$= n\frac{d^{n-1}}{dx^{n-1}}(x^{n-1})$

$= n\frac{d^{n-2}}{dx^{n-2}}\left(\frac{d}{dx}(x^{n-1})\right)$

$= n(n-1)\frac{d^{n-2}}{dx^{n-2}}(x^{n-2})$

$\vdots$

$= n!$

Exercise 12E

2 $f'(x) = \cos x + 2x$

$f'\left(\frac{\pi}{2}\right) = \cos\frac{\pi}{2} + 2 \times \frac{\pi}{2}$

$= \pi$

3 $g'(x) = \frac{1}{4\cos^2 x} + 3\sin x - 3x^2$

$g'\left(\frac{\pi}{6}\right) = \frac{1}{4\cos^2\left(\frac{\pi}{6}\right)} + 3\sin\frac{\pi}{6} - 3\left(\frac{\pi}{6}\right)^2$

$= \frac{1}{4\left(\frac{\sqrt{3}}{2}\right)^2} + 3 \times \frac{1}{2} - 3 \times \frac{\pi^2}{36}$

$= \frac{11}{6} - \frac{\pi^2}{12}$

4 $h'(x) = \cos x - \sin x$

$h'(x) = 0$

$\Rightarrow \cos x - \sin x = 0$

$\Rightarrow \tan x = 1$

$\therefore x = \frac{\pi}{4}, \frac{5\pi}{4}$

5 $\frac{dy}{dx} = 1 - \frac{2}{x^3}$

$\frac{1}{4\cos^2 x} - \frac{2}{x^3} = 1 - \frac{2}{x^3}$

$4\cos^2 x = 1$

$\cos x = \pm\frac{1}{2}$

$\therefore x = \frac{\pi}{3}, \frac{2\pi}{3}, \frac{4\pi}{3}, \frac{5\pi}{3}$

Exercise 12F

2 $f'(x) = \frac{1}{2}e^x - \frac{7}{x}$

$f'(\ln 4) = \frac{1}{2}e^{\ln 4} - \frac{7}{\ln 4}$

$= \frac{1}{2} \times 4 - \frac{7}{\ln 4} = 2 - \frac{7}{\ln 4}$

3 $f'(x) = e^x - \dfrac{1}{2x}$

$f'(\ln 3) = e^{\ln 3} - \dfrac{1}{2\ln 3}$

$= 3 - \dfrac{1}{2\ln 3} = 3 - \dfrac{1}{\ln 9}$

4 $f'(x) = -6$

$-2e^x = -6$

$e^x = 3$

$x = \ln 3$

5 $f(x) = e^x - 2x$ is increasing where $f'(x) > 0$

$f'(x) > 0$

$e^x - 2 > 0$

$e^x > 2$

$x > \ln 2$

COMMENT

Remember that because e^x and $\ln x$ are both increasing functions, taking the inverse function through an inequality is entirely valid.

6 $g'(x) = 2$

$2x - \dfrac{12}{x} = 2$

$x^2 - x - 6 = 0$

$(x+2)(x-3) = 0$

$x = -2$ or 3

However, reject $x = -2$ as not within the domain of g.

Hence $x = 3$.

COMMENT

Always check for the validity of solutions in any question containing a logarithm or square root, since the working can give rise to solution values outside the domain of the original function.

7 a i $y = 3\ln x \Rightarrow \dfrac{dy}{dx} = \dfrac{3}{x}$

  ii $y = \ln 5 + \ln x \Rightarrow \dfrac{dy}{dx} = \dfrac{1}{x}$

b i $y = e^3 \times e^x \Rightarrow \dfrac{dy}{dx} = e^3 \times e^x = e^{3+x}$

  ii $y = e^{-3} \times e^x \Rightarrow \dfrac{dy}{dx} = e^{-3} \times e^x = e^{x-3}$

c i $y = e^{\ln(x^2)} = x^2 \Rightarrow \dfrac{dy}{dx} = 2x$

  ii $y = e^2 \times e^{\ln(x^3)} = e^2 \times x^3$

  $\Rightarrow \dfrac{dy}{dx} = e^2 \times 3x^2 = 3e^2 x^2$

Exercise 12G

2 When $x = 2\ln 5 = \ln 25$,

$y = 3 - \dfrac{1}{5}e^{\ln 25}$

$= 3 - \dfrac{25}{5}$

$= -2$

$\dfrac{dy}{dx} = -\dfrac{e^x}{5}$

$\Rightarrow \dfrac{dy}{dx}(\ln 25) = -\dfrac{25}{5} = -5$

Therefore the gradient of the normal is $\dfrac{1}{5}$

The equation of the normal is

$y - y_1 = m(x - x_1)$

$y - (-2) = \dfrac{1}{5}(x - \ln 25)$

$5y + 10 = x - \ln 25$

$x - 5y - 10 - \ln 25 = 0$

3 $\dfrac{dy}{dx} = 3x^2 - 6x$

$\Rightarrow \dfrac{dy}{dx}(2) = 12 - 12 = 0$

$y(2) = 8 - 12 = -4$

∴ the equation of the tangent is $y = -4$

This intersects the curve where
$x^3 - 3x^2 = -4$
$x^3 - 3x^2 + 4 = 0$
$(x-2)^2(x+1) = 0$
$x = 2$ or -1

Thus the tangent meets the curve again at $x = -1$, at the point $(-1, -4)$.

> **COMMENT**
>
> Since there is a tangent at $x = 2$, we already know that the cubic factorises with $(x-2)^2$ as a factor (repeated root at a tangent).

4 $\dfrac{dy}{dx} = 3x^2 - 6x$

$\Rightarrow \dfrac{dy}{dx}(1) = -3$

Therefore the gradient of the normal at $(1, -2)$ is $\dfrac{1}{3}$

Require $\dfrac{dy}{dx} = \dfrac{1}{3}$

∴ $3x^2 - 6x = \dfrac{1}{3}$

$9x^2 - 18x - 1 = 0$

$x = \dfrac{18 \pm \sqrt{18^2 + 36}}{18}$

$= 1 \pm \dfrac{\sqrt{360}}{18}$

$= 1 \pm \dfrac{\sqrt{10}}{3}$

$= 2.05, -0.0541$ (3SF)

5 $\dfrac{dy}{dx} = e^x + 1$

Gradient of $y = 3x$ is 3, so require $\dfrac{dy}{dx} = 3$:

$e^x + 1 = 3$

$\Rightarrow x = \ln 2$

$y(\ln 2) = 2 + \ln 2$
$= \ln 2e^2$

Equation of the tangent is
$y - y_1 = m(x - x_1)$
$y - \ln 2e^2 = 3(x - \ln 2)$
$y = 3x - \ln 8 + \ln 2e^2$
$y = 3x + \ln\left(\dfrac{e^2}{4}\right)$
$y = 3x + 2 - \ln 4$

6 $y = (x-1)^2 = x^2 - 2x + 1$

$\Rightarrow \dfrac{dy}{dx} = 2x - 2$

Normal at the point $P(a, (a-1)^2)$ has gradient $\dfrac{-1}{2a-2}$

Equation of the normal is
$y - y_1 = m(x - x_1)$
$y - (a-1)^2 = \dfrac{-1}{2a-2}(x - a)$

Require that this passes through $(0, 0)$

∴ $0 - (a-1)^2 = \dfrac{-1}{2a-2}(0 - a)$

$-(a-1)^2 = \dfrac{a}{2(a-1)}$

$-2(a-1)^3 = a$

From GDC: $a = 0.410$ (3SF)

∴ the coordinates of P are $(0.410, 0.348)$

7

> **COMMENT**
>
> The question requires you to prove that the area is independent of a; this means that the end answer for the area should be an expression in which a does not appear. Calculate in the normal way, with the expectation that a will cancel out in the final part of the working.

$y = kx^{-1} \Rightarrow \dfrac{dy}{dx} = -kx^{-2}$

Tangent at the point (a, ka^{-1}) has gradient $m = -ka^{-2}$

Equation of the tangent is
$y - y_1 = m(x - x_1)$
$y - ka^{-1} = -ka^{-2}(x - a)$
$y = \dfrac{-k}{a^2}(x - 2a)$

This line intersects the y-axis at $P\left(0, \dfrac{2k}{a}\right)$
and intersects the x-axis at $Q(2a, 0)$

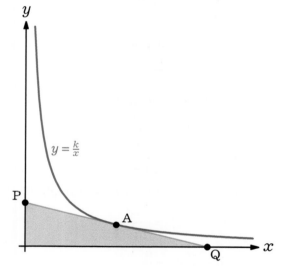

Figure 12G.7

$OP = \dfrac{2k}{a}, \quad OQ = 2a$

$\therefore$ Area $OPQ = \dfrac{1}{2}(OP)(OQ)$
$= \dfrac{1}{2} \times \dfrac{2k}{a} \times 2a$
$= 2k$

Hence the area of OPQ is independent of a.

8 $\dfrac{dy}{dx} = 3x^2 - 1$

Tangent at the point $(a, a^3 - a)$ has gradient $3a^2 - 1$

Equation of the tangent is
$y - y_1 = m(x - x_1)$
$y - (a^3 - a) = (3a^2 - 1)(x - a)$
$y = (3a^2 - 1)x - 2a^3$

This line intersects the curve where
$x^3 - x = (3a^2 - 1)x - 2a^3$
$x^3 - 3a^2 x + 2a^3 = 0$

Since the line is tangent to the curve at $x = a$, this cubic must have $(x - a)^2$ as a factor (repeated root at a tangent). Hence the cubic factorises as

$(x - a)^2 (x + 2a) = 0$

Thus the tangent intersects the curve again at $x = -2a$, as required.

Exercise 12H

2 $\dfrac{dy}{dx} = 3x^2 + 6x - 24$

Stationary points where $\dfrac{dy}{dx} = 0$:
$3x^2 + 6x - 24 = 0$
$x^2 + 2x - 8 = 0$
$(x + 4)(x - 2) = 0$
$x = -4$ or 2

$x = -4 \Rightarrow y = (-4)^3 + 3(-4)^2 - 24(-4) + 12 = 92$
$x = 2 \Rightarrow y = 2^3 + 3(2)^2 - 24(2) + 12 = -16$

$\therefore$ stationary points are $(-4, 92)$ and $(2, -16)$

$\dfrac{d^2 y}{dx^2} = 6x + 6$

$\dfrac{d^2 y}{dx^2}(-4) = -18 < 0$
$\Rightarrow (-4, 92)$ is a local maximum

$\dfrac{d^2 y}{dx^2}(2) = 18 > 0$
$\Rightarrow (2, -16)$ is a local minimum

3 $\dfrac{dy}{dx} = 1 - \dfrac{1}{2}x^{-\frac{1}{2}} = 1 - \dfrac{1}{2\sqrt{x}}$

Stationary points where $\dfrac{dy}{dx} = 0$:

$1 - \dfrac{1}{2\sqrt{x}} = 0$

$2\sqrt{x} - 1 = 0$

$\sqrt{x} = \dfrac{1}{2}$

$x = \dfrac{1}{4}$

$x = \dfrac{1}{4} \Rightarrow y = \dfrac{1}{4} - \sqrt{\dfrac{1}{4}} = -\dfrac{1}{4}$

∴ stationary point is $\left(\dfrac{1}{4}, -\dfrac{1}{4}\right)$

$\dfrac{d^2y}{dx^2} = \dfrac{1}{4}x^{-\frac{3}{2}}$

$\dfrac{d^2y}{dx^2}\left(\dfrac{1}{4}\right) = \dfrac{1}{4} \times 8 > 0$

$\Rightarrow \left(\dfrac{1}{4}, -\dfrac{1}{4}\right)$ is a local minimum

4 $\dfrac{dy}{dx} = \cos x - 4\sin x$

Stationary points where $\dfrac{dy}{dx} = 0$:

$\cos x - 4\sin x = 0$

$4\sin x = \cos x$

$\tan x = \dfrac{1}{4}$

∴ $x = 0.245$ or 3.39 (3SF)

$x = 0.245 \Rightarrow y = \sin 0.245 + 4\cos 0.245 = 4.12$

$x = 3.39 \Rightarrow y = \sin 3.39 + 4\cos 3.39 = -4.12$

∴ stationary points are $(0.245, 4.12)$ and $(3.39, -4.12)$

$\dfrac{d^2y}{dx^2} = -\sin x - 4\cos x = -y$

$\dfrac{d^2y}{dx^2}(0.245) = -4.12 < 0$

$\Rightarrow (0.245, 4.12)$ is a local maximum

$\dfrac{d^2y}{dx^2}(3.39) = 4.12 > 0$

$\Rightarrow (3.39, -4.12)$ is a local minimum

5 $f'(x) = \dfrac{1}{x} - \dfrac{k}{x^{k+1}}$

Stationary points where $f'(x) = 0$:

$\dfrac{1}{x} - \dfrac{k}{x^{k+1}} = 0$

$\dfrac{1}{x} = \dfrac{k}{x^{k+1}}$

$x^k = k$

$x = k^{\frac{1}{k}}$

$x = k^{\frac{1}{k}} \Rightarrow y = \ln k^{\frac{1}{k}} + \dfrac{1}{k}$

$= \dfrac{1}{k}\ln k + \dfrac{1}{k}$

$= \dfrac{\ln k + 1}{k}$

∴ $f(x)$ has a stationary point with y-coordinate $\dfrac{\ln k + 1}{k}$

6 $f'(x) = 12x^3 - 48x^2 + 36x$

Stationary points where $f'(x) = 0$:

$12x^3 - 48x^2 + 36x = 0$

$12x(x^2 - 4x + 3) = 0$

$x(x-1)(x-3) = 0$

$x = 0, 1$ or 3

$f(0) = 3(0)^4 - 16(0)^3 + 18(0)^2 + 6 = 6$

$f(1) = 3(1)^4 - 16(1)^3 + 18(1)^2 + 6 = 11$

$f(3) = 3(3)^4 - 16(3)^3 + 18(3)^2 + 6 = -21$

∴ stationary points are $(0, 6)$, $(1, 11)$ and $(3, -21)$

$f''(x) = 36x^2 - 96x + 36$

$f''(0) = 36 > 0 \Rightarrow (0, 6)$ is a local minimum

$f''(1) = -24 < 0$
$\Rightarrow (1, 11)$ is a local maximum

$f''(3) = 72 > 0$
$\Rightarrow (3, -21)$ is another local minimum

∴ range of f is $[-21, \infty[$

COMMENT
Instead of using second derivative analysis, it would also be valid to use knowledge of the form of a positive quartic equation to assert that the first and third stationary points must be local minima.

7 $f'(x) = e^x - 4$

Stationary points where $f'(x) = 0$:
$e^x - 4 = 0$
$x = \ln 4$
$f(\ln 4) = e^{\ln 4} - 4\ln 4 + 2 = 6 - 4\ln 4$
∴ stationary point is $(\ln 4, 6 - 4\ln 4)$
$f''(x) = e^x$
$f''(\ln 4) = 4 > 0$
$\Rightarrow (\ln 4, 6 - 4\ln 4)$ is a local minimum
∴ range of f is $[6 - \ln 4, \infty[$

8 $\dfrac{dy}{dx} = 3kx^2 + 12x$

Stationary points where $\dfrac{dy}{dx} = 0$:
$3kx^2 + 12x = 0$
$kx^2 + 4x = 0$
$x(kx + 4) = 0$
$x = 0$ or $-\dfrac{4}{k}$

$x = 0 \Rightarrow y = k(0)^3 + 6(0)^2 = 0$

$x = -\dfrac{4}{k} \Rightarrow y = k\left(-\dfrac{4}{k}\right)^3 + 6\left(-\dfrac{4}{k}\right)^2 = \dfrac{32}{k^2}$

∴ stationary points are $(0, 0)$ and $\left(-\dfrac{4}{k}, \dfrac{32}{k^2}\right)$

$\dfrac{d^2y}{dx^2} = 6kx + 12$

$\dfrac{d^2y}{dx^2}(0) = 12 > 0$
$\Rightarrow (0, 0)$ is a local minimum

$\dfrac{d^2y}{dx^2}\left(-\dfrac{4}{k}\right) = -12 < 0$
$\Rightarrow \left(-\dfrac{4}{k}, \dfrac{32}{k^2}\right)$ is a local maximum

Exercise 12I

1 $\dfrac{dy}{dx} = e^x - 2x$

$\dfrac{d^2y}{dx^2} = e^x - 2$

Points of inflexion where $\dfrac{d^2y}{dx^2} = 0$:

$e^x - 2 = 0$
$x = \ln 2$
$x = \ln 2 \Rightarrow y = e^{\ln 2} - (\ln 2)^2 = 2 - (\ln 2)^2$
∴ point of inflexion is at $\left(\ln 2, 2 - (\ln 2)^2\right)$

2 $\dfrac{dy}{dx} = 4x^3 - 12x + 7$

$\dfrac{d^2y}{dx^2} = 12x^2 - 12$

Points of inflexion where $\dfrac{d^2y}{dx^2} = 0$:

$12x^2 - 12 = 0$
$x^2 = 1$
$x = \pm 1$

$x = 1 \Rightarrow y = 1^4 - 6(1)^2 + 7(1) + 2 = 4$

$x = -1 \Rightarrow y = (-1)^4 - 6(-1)^2 + 7(-1) + 2 = -10$

∴ points of inflexion are at $(1, 4)$ and $(-1, -10)$

3 $\dfrac{dy}{dx} = \cos x$

$\dfrac{d^2y}{dx^2} = -\sin x = -y$

Points of inflexion have $\dfrac{d^2y}{dx^2} = 0$, which must therefore be on $y = 0$, the x-axis.

4 $\dfrac{dy}{dx} = -2\sin x + 1$

$\dfrac{d^2y}{dx^2} = -2\cos x$

Points of inflexion where $\dfrac{d^2y}{dx^2} = 0$:

$-2\cos x = 0$

$\cos x = 0$

$\therefore x = \dfrac{\pi}{2}, \dfrac{3\pi}{2}$

$x = \dfrac{\pi}{2} \Rightarrow y = 2\cos\left(\dfrac{\pi}{2}\right) + \dfrac{\pi}{2} = \dfrac{\pi}{2}$

$x = \dfrac{3\pi}{2} \Rightarrow y = 2\cos\left(\dfrac{3\pi}{2}\right) + \dfrac{3\pi}{2} = \dfrac{3\pi}{2}$

Verifying that these are points of inflexion:

For a small positive value δ,

$\cos\left(\dfrac{\pi}{2} - \delta\right) > 0$ and so $\dfrac{d^2y}{dx^2}\left(\dfrac{\pi}{2} - \delta\right) < 0$

(gradient of curve is decreasing);

similarly, $\cos\left(\dfrac{\pi}{2} + \delta\right) < 0$ and so

$\dfrac{d^2y}{dx^2}\left(\dfrac{\pi}{2} + \delta\right) > 0$ (gradient of curve is

increasing). Therefore $\left(\dfrac{\pi}{2}, \dfrac{\pi}{2}\right)$ is a genuine point of inflexion.

For a small positive value δ,

$\cos\left(\dfrac{3\pi}{2} - \delta\right) < 0$ and so $\dfrac{d^2y}{dx^2}\left(\dfrac{3\pi}{2} - \delta\right) > 0$

(gradient increasing); similarly,

$\cos\left(\dfrac{3\pi}{2} + \delta\right) > 0$ and so $\dfrac{d^2y}{dx^2}\left(\dfrac{3\pi}{2} + \delta\right) < 0$

(gradient decreasing). Therefore $\left(\dfrac{3\pi}{2}, \dfrac{3\pi}{2}\right)$

is a genuine point of inflexion.

> **COMMENT**
> Remember that just showing that the second derivative is zero is not sufficient for the point to be an inflexion. Further working is needed, either checking values on either side, as above, or by showing that the first non-zero derivative after the second derivative is odd. An alternative to the justifications using δ would be:
>
> $\dfrac{d^3y}{dx^3} = 2\sin x$
>
> $\dfrac{d^3y}{dx^3}\left(\dfrac{\pi}{2}\right) = 2 \neq 0$
>
> $\Rightarrow \left(\dfrac{\pi}{2}, \dfrac{\pi}{2}\right)$ is a genuine point of inflexion
>
> $\dfrac{d^3y}{dx^3}\left(\dfrac{3\pi}{2}\right) = -2 \neq 0$
>
> $\Rightarrow \left(\dfrac{3\pi}{2}, \dfrac{3\pi}{2}\right)$ is a genuine point of inflexion

5 $\dfrac{dy}{dx} = 3x^2 - 2ax - b$

$\dfrac{d^2y}{dx^2} = 6x - 2a$

Points of inflexion where $\dfrac{d^2y}{dx^2} = 0$:

$6x - 2a = 0$

$x = \dfrac{a}{3}$

If this is also to be a stationary point, then require $\dfrac{dy}{dx}\left(\dfrac{a}{3}\right) = 0$:

$3\left(\dfrac{a}{3}\right)^2 - 2a\left(\dfrac{a}{3}\right) - b = 0$

$\dfrac{a^2}{3} - \dfrac{2a^2}{3} - b = 0$

$\Rightarrow b = -\dfrac{a^2}{3}$

12 Basic differentiation and its applications

> **COMMENT**
>
> Only positive cubics with a horizontal inflexion point and leading coefficient 1 have the form $y = (x-k)^3 + d$, which expands to $y = x^3 - 3kx^2 + 3k^2x - k^3 + d$. Comparing this form with the equation in the question gives $a = 3k$ and $b = -3k^2 = -\dfrac{a^2}{3}$, as required.

6 Graph shows the gradient function.

When $f'(x) = 0$ there is a stationary point.

If at a stationary point the gradient changes from negative to positive, it is a local minimum (A).

If at a stationary point the gradient changes from positive to negative, it is a local maximum (B).

When $f'(x)$ is itself at a local maximum or minimum, $f''(x) = 0$ and there is a point of inflexion (C).

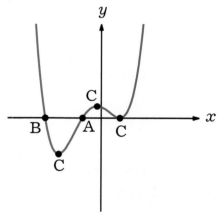

Figure 12I.6

Exercise 12J

1 $\dfrac{d}{dx}(e^x) = e^x$

$\dfrac{d}{dx}(e^x) \neq 0$ for all x, so there are no stationary points; the end points $x = 0$ and $x = 1$ must give the minimum and maximum for the interval.

$e^0 = 1$ is the minimum

$e^1 = e$ is the maximum

2 a Area $A = x(30-x) = 30x - x^2$

$\dfrac{dA}{dx} = 30 - 2x$

Stationary point when $\dfrac{dA}{dx} = 0$:

$30 - 2x = 0$

$x = 15$

Since the extreme values $x = 0$ and $x = 30$ clearly give zero area while intermediate values give a positive area, the end points do not provide a maximum value, and the stationary point is a maximum.

∴ maximum area $= 15 \times 15 = 225 \text{ m}^2$

b Perimeter $= 2x + 2(30 - x)$
$= 60 \text{ m}$, a constant

3 $\dfrac{dy}{dx} = 3x^2 - 9$

Stationary points where $\dfrac{dy}{dx} = 0$:

$3x^2 - 9 = 0$

$x^2 = 3$

$x = \pm\sqrt{3}$

$x = -\sqrt{3} \Rightarrow y = (-\sqrt{3})^3 - 9(-\sqrt{3}) = 6\sqrt{3}$

$x = \sqrt{3} \Rightarrow y = (\sqrt{3})^3 - 9(\sqrt{3}) = -6\sqrt{3}$

Checking the end points:

$x = -2 \Rightarrow y = (-2)^3 - 9(-2) = 10 > -6\sqrt{3}$

$x = 5 \Rightarrow y = 5^3 - 9 \times 5 = 80 > 6\sqrt{3}$

The minimum value over the interval $[-2, 5]$ is $-6\sqrt{3} = -10.4$

The maximum value over the interval $[-2, 5]$ is 80

4 $f'(x) = e^x - 3$

Stationary points when $\dfrac{dy}{dx} = 0$:

$e^x - 3 = 0$

$x = \ln 3$

$x = \ln 3 \Rightarrow y = 3 - 3\ln 3 = -0.296$

Checking the end points:

$x = 0 \Rightarrow y = 1$

$x = 2 \Rightarrow y = e^2 - 6 = 1.39$

The minimum value over the interval $[0, 2]$ is $3 - 3\ln 3 = -0.296$

The maximum value over the interval $[0, 2]$ is $e^2 - 6 = 1.39$

5 $\dfrac{dy}{dx} = \cos x + 2$

Stationary points when $\dfrac{dy}{dx} = 0$:

$\cos x + 2 = 0$

$\cos x = -2$

No solutions so no stationary points.

Checking the end points:

$x = 0 \Rightarrow y = 0$

$x = 2\pi \Rightarrow y = 4\pi$

The minimum value over the interval $[0, 2\pi]$ is 0

The maximum value over the interval $[0, 2\pi]$ is 4π

6 $f(x) = x + \dfrac{1}{x}$ for $x > 0$

$f'(x) = 1 - \dfrac{1}{x^2}$

Stationary points where $f'(x) = 0$:

$1 - \dfrac{1}{x^2} = 0$

$x^2 = 1$

$\therefore x = 1$ (as $x > 0$)

Classify the stationary point:

$f''(x) = 2x^{-3}$

$f''(1) = 2 > 0 \Rightarrow$ local minimum

$\therefore$ minimum value of f is 2

7 Distance $D = \left(1 - w^{-\frac{1}{2}}\right) \times \dfrac{5}{w} = 5\left(w^{-1} - w^{-\frac{3}{2}}\right)$

$\dfrac{dD}{dw} = 5\left(\dfrac{3}{2}w^{-\frac{5}{2}} - w^{-2}\right)$

Stationary points when $\dfrac{dD}{dw} = 0$:

$5\left(\dfrac{3}{2}w^{-\frac{5}{2}} - w^{-2}\right) = 0$

$\dfrac{3}{2}w^{-\frac{5}{2}} = w^{-2}$

$\dfrac{3}{2} = \sqrt{w}$

$w = \dfrac{9}{4}$

Classify the stationary point:

$\dfrac{d^2D}{dw^2} = 5\left(2w^{-3} - \dfrac{15}{4}w^{-\frac{7}{2}}\right)$

$\dfrac{d^2D}{dw^2}\left(\dfrac{9}{4}\right) = -0.219 < 0 \Rightarrow$ local maximum.

So a weight of $\dfrac{9}{4} = 2.25$ will maximise the distance travelled.

8 $\dfrac{dt}{dp} = \dfrac{p}{5000} + \dfrac{1}{100} = \dfrac{p + 50}{5000}$

Stationary points when $\dfrac{dt}{dp} = 0$:

$\dfrac{p + 50}{5000} = 0$

$\Rightarrow p = -50 \notin [0, 100]$

So there are no stationary points in the interval.

Checking the end points:

$p = 0 \Rightarrow t = 2$ minutes, the minimum time to melt $100\,\text{g}$

$p = 100 \Rightarrow t = 4$ minutes, the maximum time to melt $100\,\text{g}$

9 a $-1 \leq \cos t \leq 1$

$\therefore$ V has a range of $[40, 160]$

So the minimum volume is 40 million litres.

b Water flow will equal the rate of change in volume:

$$\text{Flow} = \frac{dV}{dt} = -60\sin t$$

Maximum flow occurs when $\frac{d\,\text{Flow}}{dt} = 0$:

$-60\cos t = 0$

$\cos t = 0$

$t = \frac{\pi}{2}, \frac{3\pi}{2} \quad (t \in [0,6])$

So the maximum flow in the first 6 days occurs at $t = \frac{\pi}{2}$ (1.6 days) and $\frac{3\pi}{2}$ (4.7 days).

> **COMMENT**
>
> Although $t = \frac{\pi}{2}$ is actually a local minimum of the 'Flow' function, it still represents a maximum flow of water: the negative sign of 'Flow' when $t = \frac{\pi}{2}$ (which makes it a minimum in the sense of being at its most negative) just means that water is flowing out at that point and not in, but the rate at which the water is flowing is exactly the same as at the local maximum when $t = \frac{3\pi}{2}$ (both are 60 million litres per day).

10 a $\frac{dF}{ds} = 4 - 2s$

Maximum F occurs when $\frac{dF}{ds} = 0$ or at an end point of the domain [0, 4.2]

$\frac{dF}{ds} = 0$

$4 - 2s = 0$

$\Rightarrow s = 2$

and $F(2) = 5$

Check end points:
$F(0) = 1, F(4.2) = 0.16$

$\therefore$ maximum F occurs at $s = 2$

b Minimum C occurs when $\frac{dC}{ds} = 0$ or at an end point of the domain.

$\frac{dC}{ds} = 0$

$0.2(4 - 2s) + 0.1 = 0$

$0.9 - 0.4s = 0$

$\Rightarrow s = \frac{9}{4}$

and $C\left(\frac{9}{4}\right) = 1.5125$

Check end points:
$C(0) = 0.5, C(4.2) = 0.752$

$\therefore$ minimum C occurs when $s = 0$

c Profit P is given by

$P = F - C$

$= F - (0.3 + 0.2F + 0.1s)$

$= 0.8F - 0.3 - 0.1s$

$= 0.8(4s + 1 - s^2) - 0.3 - 0.1s$

$= -0.8s^2 + 3.1s + 0.5$

Stationary point of P occurs when $\frac{dP}{ds} = 0$:

$-1.6s + 3.1 = 0$

$\Rightarrow s = \frac{31}{16} = 1.94 \text{ (3SF)}$

Since this value of s lies inside the domain [0, 4.2] and is the position of the vertex of a negative quadratic, it must give the global maximum of P over the domain.

> **COMMENT**
>
> In part (b) we can actually avoid the calculus altogether, because the minimum value for a negative quadratic over a restricted domain must lie at one of the end points; it cannot be the stationary point, since that must be a maximum. If you prefer to use this argument in an examination, be explicit about your reasoning.

11 a $V(0) = 4$, so 4 litres of petrol was initially in the tank.

b 30 seconds = 0.5 minutes
$V(0.5) = 41.5$, so the capacity of the tank is 41.5 litres.

c Flow $= \dfrac{dV}{dt} = 600t - 900t^2$

Maximum flow when $\dfrac{d\,\text{Flow}}{dt} = 0$:

$600 - 1800t = 0$

$t = \dfrac{1}{3}$

∴ maximum flow is at 20 seconds.

12 a Total energy $E = x\left(2 - \dfrac{x}{10}\right)$

$= 2x - \dfrac{x^2}{10}\,\text{kJ}$

b Maximum energy when $\dfrac{dE}{dt} = 0$:

$2 - \dfrac{2x}{10} = 0$

$x = 10$

∴ a total surface area of 10 m² provides maximum energy.

c Net energy
$N = E - 0.01x^3 = 2x - \dfrac{x^2}{10} - 0.01x^3$

Leaves produce more energy than they require when $N > 0$:

$2x - \dfrac{x^2}{10} - 0.01x^3 > 0$

$\dfrac{1}{100}(200x - 10x^2 - x^3) > 0$

$x(200 - 10x - x^2) > 0$

$x(20 + x)(10 - x) > 0$

$\Rightarrow x \in \,]0, 10[$

d Maximum net energy when $\dfrac{dN}{dt} = 0$:

$2 - \dfrac{x}{5} - 0.03x^2 = 0$

$\dfrac{1}{100}(200 - 20x - 3x^2) = 0$

$3x^2 + 20x - 200 = 0$

$x = \dfrac{-20 \pm \sqrt{20^2 + 2400}}{6}$

Require the positive solution

$\therefore x = \dfrac{-20 + \sqrt{2800}}{6} = \dfrac{10(-1 + \sqrt{7})}{3}$

Mixed examination practice 12

Short questions

1 $\dfrac{dy}{dx} = e^x + 2\cos x$

$\dfrac{dy}{dx}\left(\dfrac{\pi}{2}\right) = e^{\frac{\pi}{2}}$

$y\left(\dfrac{\pi}{2}\right) = e^{\frac{\pi}{2}} + 2$

The equation of the tangent is

$y - y_1 = m(x - x_1)$

$y - \left(e^{\frac{\pi}{2}} + 2\right) = e^{\frac{\pi}{2}}\left(x - \dfrac{\pi}{2}\right)$

$y = 2 + e^{\frac{\pi}{2}}\left(x + 1 - \dfrac{\pi}{2}\right)$

2 $y = x^3 - 6x^2 + 12x - 8$

$\Rightarrow \dfrac{dy}{dx} = 3x^2 - 12x + 12 = 3(x - 2)^2$

At $x = 2$, $\dfrac{dy}{dx} = 0$ and $y = 0$, so the normal is a vertical line through (2, 0)

i.e. its equation is $x = 2$

12 Basic differentiation and its applications

3 $f(1) = 2$
$\Rightarrow 1 + b + c = 2$
$\Rightarrow b + c = 1$
$f'(x) = 2x + b$
$f'(2) = 12$
$\Rightarrow 4 + b = 12$
$\Rightarrow b = 8$
$\therefore c = 1 - 8 = -7$

4 $\dfrac{dy}{dx} = \dfrac{1}{2\sqrt{x}} + 3$

$\dfrac{dy}{dx} = 5$

$\Rightarrow \dfrac{1}{2\sqrt{x}} + 3 = 5$

$\dfrac{1}{2\sqrt{x}} = 2$

$\sqrt{x} = \dfrac{1}{4}$

$\Rightarrow x = \dfrac{1}{16}$

$y\left(\dfrac{1}{16}\right) = \sqrt{\dfrac{1}{16}} + 3\left(\dfrac{1}{16}\right) = \dfrac{7}{16}$

$\therefore$ the coordinates are $\left(\dfrac{1}{16}, \dfrac{7}{16}\right)$

5 $y' = \dfrac{x^2}{2} - 2x + 1$

$y'' = x - 2$

Inflexion where $y'' = 0$
i.e. $x = 2$

$y(2) = \dfrac{2^3}{6} - 2^2 + 2 = -\dfrac{2}{3}$

$\therefore$ the coordinates are $\left(2, -\dfrac{2}{3}\right)$

6 $\dfrac{dy}{dx} = \sec^2 x - \dfrac{4}{3}$

Stationary points where $\dfrac{dy}{dx} = 0$:

$\sec^2 x - \dfrac{4}{3} = 0$

$\cos^2 x = \dfrac{3}{4}$

$\cos x = \pm\dfrac{\sqrt{3}}{2}$

$\therefore x = [2n\pi +]\dfrac{\pi}{6}, \dfrac{5\pi}{6}, \dfrac{7\pi}{6}, \dfrac{11\pi}{6}$ ($n \in \mathbb{Z}$)

$\dfrac{d^2y}{dx^2} = 2\sec^2 x \tan x$

(so the sign of $\dfrac{d^2y}{dx^2}$ is determined by the sign of $\tan x$)

At $x = [2n\pi +]\dfrac{\pi}{6}, \dfrac{7\pi}{6}$, $\dfrac{d^2y}{dx^2} > 0$
$\Rightarrow$ local minima

At $x = [2n\pi +]\dfrac{5\pi}{6}, \dfrac{11\pi}{6}$, $\dfrac{d^2y}{dx^2} < 0$
$\Rightarrow$ local maxima

So there are local maxima at $x = \dfrac{5\pi}{6} + n\pi$
and local minima at $x = \dfrac{\pi}{6} + n\pi$ ($n \in \mathbb{Z}$).

7 a Local minimum: $f'(x) = 0$ and gradient on graph of $f'(x)$ is positive (A)

b Local maximum: $f'(x) = 0$ and gradient on graph of $f'(x)$ is negative (B)

c Inflexion: turning points on graph of $f'(x)$ (C)

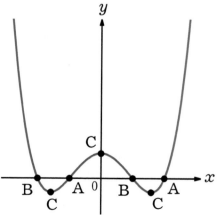

Figure 12MS.7

132 Mixed examination practice 12

8 $\dfrac{dy}{dx} = 3x^2$

Tangent at (a, a^3) has gradient $3a^2$ and so has equation

$y - a^3 = 3a^2(x-a)$

$y = 3a^2 x - 2a^3$

It has y-intercept at $(0, -2a^3)$

Gradient of curve at $(-a, -a^3)$ is $3a^2$

Normal at $(-a, -a^3)$ has gradient $-\dfrac{1}{3a^2}$ and so has equation

$y - (-a^3) = -\dfrac{1}{3a^2}(x+a)$

$y = -\dfrac{1}{3a^2}x - a^3 - \dfrac{1}{3a}$

It has y-intercept at $\left(0, -a^3 - \dfrac{1}{3a}\right)$

If the y-intercepts are the same point, then

$-2a^3 = -a^3 - \dfrac{1}{3a}$

$\dfrac{1}{3a} = a^3$

$3a^4 = 1$

$a = \pm 3^{-\tfrac{1}{4}} = \pm \dfrac{1}{\sqrt[4]{3}}$

Long questions

1 a Point of contact at $x = 2$

On the tangent line, $y = 24(2-1) = 24$

On the curve,

$y = a(2)^3 + b(2)^2 + 4$

$= 8a + 4b + 4$

$\therefore 8a + 4b + 4 = 24$

$\Rightarrow 2a + b = 5$

b For the curve, $\dfrac{dy}{dx} = 3ax^2 + 2bx$

At $x = 2$,

$\dfrac{dy}{dx} = 3a(2)^2 + 2b(2)$

$= 12a + 4b$

and gradient of the tangent is 24

$\therefore 12a + 4b = 24$

$\Rightarrow 3a + b = 6$

c $2a + b = 5$ ...(1)

$3a + b = 6$ ...(2)

$(2) - (1) \Rightarrow a = 1$

$\therefore b = 3$

d Points of intersection occur when

$x^3 + 3x^2 + 4 = 24(x-1)$

$x^3 + 3x^2 - 24x + 28 = 0$

One solution is known to be at $x = 2$, which is a double root.

So $(x-2)^2$ is a factor of $x^3 + 3x^2 - 24x + 28$; factorising gives

$(x-2)^2(x+7) = 0$

$\therefore$ the other intersection point is $(x, y) = (-7, -192)$

2 a i $\dfrac{dy}{dx} = 3x^2 - 2x - 1$

Stationary points (including the turning point at A) occur where $\dfrac{dy}{dx} = 0$:

$3x^2 - 2x - 1 = 0$

$(3x+1)(x-1) = 0$

$x = -\dfrac{1}{3}$ or 1

The point A has negative x-coordinate, so $A = \left(-\dfrac{1}{3}, \dfrac{86}{27}\right)$

ii $\dfrac{d^2 y}{dx^2} = 6x - 2$

Points of inflexion occur where $\dfrac{d^2 y}{dx^2} = 0$:

$6x - 2 = 0$

12 Basic differentiation and its applications

$$\Rightarrow x = \frac{1}{3}$$

∴ the coordinates of B are $\left(\frac{1}{3}, \frac{70}{27}\right)$

b i The line containing A and B has gradient

$$m = \frac{\frac{86}{27} - \frac{70}{27}}{-\frac{1}{3} - \frac{1}{3}}$$

$$= -\frac{\frac{16}{27}}{\frac{2}{3}}$$

$$= -\frac{8}{9}$$

The equation of the line is

$$y - y_1 = m(x - x_1)$$

$$y - \frac{70}{27} = -\frac{8}{9}\left(x - \frac{1}{3}\right)$$

$$27y - 70 = -24x + 8$$

$$27y + 24x = 78 \quad \text{or} \quad y = -\frac{8}{9}x + \frac{78}{27}$$

ii Require gradient of tangent to be $-\frac{8}{9}$

i.e. $\frac{dy}{dx} = -\frac{8}{9}$

$$3x^2 - 2x - 1 = -\frac{8}{9}$$

$$27x^2 - 18x - 1 = 0$$

$$x = \frac{18 \pm \sqrt{324 + 108}}{54}$$

$$= \frac{18 \pm 12\sqrt{3}}{54}$$

$$= \frac{3 \pm 2\sqrt{3}}{9}$$

$$y\left(\frac{3\pm 2\sqrt{3}}{9}\right) = \left(\frac{3\pm 2\sqrt{3}}{9}\right)^3 - \left(\frac{3\pm 2\sqrt{3}}{9}\right)^2 - \left(\frac{3\pm 2\sqrt{3}}{9}\right) + 3$$

$$= \frac{27 \pm 54\sqrt{3} + 108 \pm 24\sqrt{3}}{729} - \frac{9 \pm 12\sqrt{3} + 12}{81} - \frac{3 \pm 2\sqrt{3}}{9} + 3$$

$$= \frac{1}{243}(45 \pm 26\sqrt{3} - 63 \mp 36\sqrt{3} - 81 \mp 54\sqrt{3} + 729)$$

$$= \frac{70}{27} \mp \frac{63\sqrt{3}}{243}$$

∴ the tangent lines are

$$y = -\frac{8}{9}\left(x - \frac{3 \pm 2\sqrt{3}}{9}\right) + \frac{70}{27} \mp \frac{64\sqrt{3}}{243}$$

3 a i $P(0) = 10 + 1 - 0 = 11$, so the initial population is 11 000.

ii 14 million = 14 000 thousand

$P = 14000$

$10 + e^t - 3t = 14000$

From GDC, $t = 9.55$ (3SF)

So after 9.55 hours the population reaches 14 million.

b i $\frac{dP}{dt} = e^t - 3$

ii 6 million = 6000 thousand

$\frac{dP}{dt} = 6000$

$e^t - 3 = 6000$

$e^t = 6003$

$t = \ln 6003$

$= 8.70$ hours (3SF)

c i $\frac{d^2P}{dt^2} = e^t$, the rate of acceleration of the population

ii $\frac{dP}{dt} = 0$

$\Rightarrow e^t - 3 = 0$

$\Rightarrow t = \ln 3$

At $t = \ln 3$, $\frac{d^2P}{dt^2} = 3 > 0$, so this is a local minimum.

$P(\ln 3) = 10 + 3 - 3\ln 3 = 9.704$

∴ the minimum population is 9704.

13 Basic integration and its applications

Exercise 13B

1 For any 3 values of $c \in \mathbb{R}$:

a $\dfrac{3x^4}{4}+c$

b c

Exercise 13C

4 $\displaystyle\int \dfrac{1+x}{\sqrt{x}}\,dx = \int x^{-\frac{1}{2}}+x^{\frac{1}{2}}\,dx$

$= 2x^{\frac{1}{2}} + \dfrac{2}{3}x^{\frac{3}{2}}+c$

$= \dfrac{2\sqrt{x}}{3}(3+x)+c$

> **COMMENT**
> After studying Section 15C, try performing this integration using a substitution $u=\sqrt{x}$.

Exercise 13E

2 $\displaystyle\int \pi(\cos x - 1)\,dx = \pi(\sin x - x)+c$

$= \pi\sin x - \pi x + c$

3

$\displaystyle\int \dfrac{\cos 2x}{\cos x - \sin x}\,dx = \int \dfrac{\cos^2 x - \sin^2 x}{\cos x - \sin x}\,dx$

$= \displaystyle\int \dfrac{(\cos x + \sin x)(\cos x - \sin x)}{\cos x - \sin x}\,dx$

$= \displaystyle\int \cos x + \sin x\,dx$

$= \sin x - \cos x + c$

Exercise 13F

2 a $f'(x) = \dfrac{1}{2x}$

$f(x) = \displaystyle\int \dfrac{1}{2x}\,dx$

$= \dfrac{1}{2}\ln x + c$

$= \ln\sqrt{x}+c$

> **COMMENT**
> An equivalent solution is $f(x) = \ln(k\sqrt{x})$ where the unknown $k = e^c$ is restricted to a positive value. Unless the question requires it, or if doing so simplifies the appearance of the equation, there is no need to rewrite logarithm solutions in this way.

b $f(2) = 7$

$\Rightarrow \dfrac{1}{2}\ln 2 + c = 7$

$\Rightarrow c = 7 - \dfrac{1}{2}\ln 2$

$\therefore f(x) = \dfrac{1}{2}\ln x + 7 - \dfrac{1}{2}\ln 2$

$= 7 + \ln\sqrt{\dfrac{x}{2}}$

3 a Maximum occurs where $\dfrac{dy}{dx} = 0$

i.e. $x^2 - 4 = 0$

$x = \pm 2$

$\dfrac{d^2y}{dx^2} = 2x$

$\dfrac{d^2y}{dx^2}(2) = 4 > 0 \Rightarrow$ local minimum

$\dfrac{d^2y}{dx^2}(-2) = -4 < 0 \Rightarrow$ local maximum

∴ maximum point occurs at $x = -2$

b $y = \int x^2 - 4 \, dx$

$= \dfrac{x^3}{3} - 4x + c$

$y(0) = 2$

$\Rightarrow \dfrac{0^3}{3} - 4(0) + c = 2$

$\Rightarrow c = 2$

∴ $y(x) = \dfrac{x^3}{3} - 4x + 2$

Hence $y(-2) = -\dfrac{8}{3} + 8 + 2 = 7\dfrac{1}{3}$

4 Gradient of normal is $x \Rightarrow$ gradient of tangent is $-\dfrac{1}{x}$

$\dfrac{dy}{dx} = -\dfrac{1}{x}$

$\Rightarrow y = \int -\dfrac{1}{x} \, dx$

$= -\ln x + c$

$y(e^2) = 3$

$\Rightarrow -2 + c = 3$

$\Rightarrow c = 5$

∴ $y = 5 - \ln x = \ln\left(\dfrac{e^5}{x}\right)$

Exercise 13G

3 $\displaystyle\int_0^\pi e^x + \sin x + 1 \, dx = \left[e^x - \cos x + x\right]_0^\pi$

$= \left(e^\pi - (-1) + \pi\right) - (1 - 1 + 0)$

$= e^\pi + 1 + \pi$

4 $\displaystyle\int_k^{2k} \dfrac{1}{x} \, dx = \left[\ln x\right]_k^{2k}$

$= \ln 2k - \ln k$

$= \ln\left(\dfrac{2k}{k}\right)$

$= \ln 2$

and this is independent of k.

5 $\displaystyle\int_3^9 2f(x) + 1 \, dx = 2\int_3^9 f(x) \, dx + \int_3^9 1 \, dx$

$= 2 \times 7 + [x]_3^9$

$= 14 + (9 - 3)$

$= 20$

6 $\displaystyle\int_1^a t^{\frac{1}{2}} \, dt = 42$

$\left[\dfrac{2}{3} t^{\frac{3}{2}}\right]_1^a = 42$

$\dfrac{2}{3} a^{\frac{3}{2}} - \dfrac{2}{3} = 42$

$a^{\frac{3}{2}} - 1 = 63$

$a = 64^{\frac{2}{3}}$

$a = 16$

Exercise 13H

2

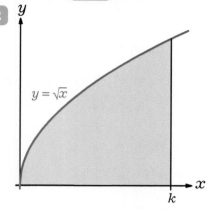

Figure 13H.2 Area enclosed by the curve $y = \sqrt{x}$, the x-axis and the line $x = k$

$$\int_0^k x^{\frac{1}{2}} \, dx = 18$$

$$\left[\frac{2}{3} x^{\frac{3}{2}}\right]_0^k = 18$$

$$\frac{2}{3} k^{\frac{3}{2}} = 18$$

$$k = 27^{\frac{2}{3}}$$

$$k = 9$$

3 a $\int_0^3 x^2 - 1 \, dx = \left[\frac{x^3}{3} - x\right]_0^3$

$$= (9 - 3) - 0$$

$$= 6$$

b

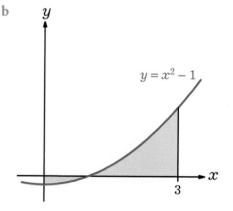

Figure 13H.3 Area bounded by the curve $y = x^2 - 1$ and the x-axis between $x = 0$ and $x = 3$

Intersections of $y = x^2 - 1$ with the x-axis occur at $x = \pm 1$.

$$\text{Area} = \int_0^3 |x^2 - 1| \, dx$$

$$= \int_0^1 1 - x^2 \, dx + \int_1^3 x^2 - 1 \, dx$$

$$= \left[x - \frac{x^3}{3}\right]_0^1 + \left[\frac{x^3}{3} - x\right]_1^3$$

$$= \left(1 - \frac{1}{3}\right) - 0 + (9 - 3) - \left(\frac{1}{3} - 1\right)$$

$$= 8 - \frac{2}{3}$$

$$= 7\frac{1}{3}$$

> **COMMENT**
>
> Alternatively, use a GDC to calculate the integral of the modulus function. Unless the question explicitly calls for an 'exact' solution, this is often a faster way of finding the solution.

4

> **COMMENT**
>
> If the area above the x-axis equals the area below it, then the net area will equal zero, i.e. the integral is zero. Using this fact is much simpler than splitting the integral into two parts and equating them.

Require that the net area equals zero:

$$\int_0^3 x^2 - kx \, dx = 0$$

$$\left[\frac{x^3}{3} - \frac{kx^2}{2}\right]_0^3 = 0$$

$$9 - \frac{9k}{2} = 0$$

$$k = 2$$

5 Intersections with the x-axis occur where

$7x - x^2 - 10 = 0$

$x^2 - 7x + 10 = 0$

$(x-2)(x-5) = 0$

$x = 2$ or 5

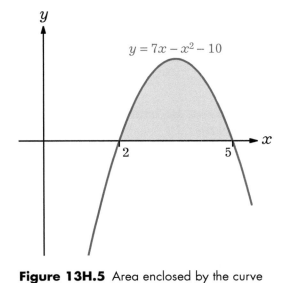

Figure 13H.5 Area enclosed by the curve $y = 7x - x^2 - 10$ and the x-axis

Hence the enclosed area is given by

$$\int_2^5 7x - x^2 - 10 \, dx = \left[\frac{7x^2}{2} - \frac{x^3}{3} - 10x \right]_2^5$$

$$= \left(\frac{175}{2} - \frac{125}{3} - 50 \right) - \left(\frac{28}{2} - \frac{8}{3} - 20 \right)$$

$$= \frac{-25}{6} - \left(-\frac{52}{6} \right)$$

$$= \frac{27}{6} = \frac{9}{2}$$

COMMENT

Once the limits are established, the integral could alternatively be calculated using a GDC.

Exercise 13I

2 Intersections when

$x^2 + x - 2 = x + 2$

$x^2 - 4 = 0$

$x = \pm 2$

Enclosed area $= \int_{-2}^{2} 4 - x^2 \, dx$

$$= \left[4x - \frac{x^3}{3} \right]_{-2}^{2}$$

$$= \left(8 - \frac{8}{3} \right) - \left(-8 + \frac{8}{3} \right)$$

$$= 16 - \frac{16}{3}$$

$$= 10\frac{2}{3}$$

3 Intersections when $e^x = x^2$

From GDC: $x = -0.703$, which lies outside the interval of interest, $[0, 2]$.

At $x = 0$, $e^x > x^2$, so $e^x - x^2$ is positive over the whole of the interval of interest.

Enclosed area $= \int_0^2 e^x - x^2 \, dx$

$$= \left[e^x - \frac{x^3}{3} \right]_0^2$$

$$= \left(e^2 - \frac{8}{3} \right) - 1$$

$$= e^2 - \frac{11}{3}$$

COMMENT

Instead of checking for roots to $e^x = x^2$, this question could be answered by using a GDC to calculate $\int_0^2 |e^x - x^2| \, dx$ directly.

13 Basic integration and its applications

4 Intersections when $\frac{1}{x} = \sin x$

From GDC: $x = 1.11$ or 2.77 for $x \in \,]0, \pi[$

Enclosed area $= \int_{1.11}^{2.77} \sin x - \frac{1}{x} dx$

$= 0.462$ (3SF)

5 The y-coordinates of the intersections are:

at $x = -1$, $y = (-1)^2 = 1$

at $x = 2$, $y = 2^2 = 4$

i.e. the intersection points are $(-1, 1)$ and $(2, 4)$.

Gradient of line $= \frac{4-1}{2-(-1)} = 1$

∴ equation of the line is

$y - y_1 = m(x - x_1)$

$y - 1 = 1(x - (-1))$

$y = x + 2$

The shaded region is the area between $y = x^2$ and $y = x + 2$:

Shaded area $= \int_{-1}^{2} x + 2 - x^2 dx$

$= \left[\frac{x^2}{2} + 2x - \frac{x^3}{3} \right]_{-1}^{2}$

$= \left(\frac{4}{2} + 4 - \frac{8}{3} \right) - \left(\frac{1}{2} - 2 + \frac{1}{3} \right)$

$= \frac{9}{2}$

> **COMMENT**
>
> An alternative approach would be to integrate $y = x^2$ between 1 and 2, and subtract that result from the area of the trapezium. As long as your working is clearly laid out, any valid method is acceptable.

6 Intersection occurs at $x = \frac{\pi}{4}$

(by symmetry or by solving $\sin x = \cos x$, i.e. $\tan x = 1$)

Shaded area $= \int_{0}^{\pi/4} \sin x \, dx + \int_{\pi/4}^{\pi/2} \cos x \, dx$

$= [-\cos x]_0^{\pi/4} + [\sin x]_{\pi/4}^{\pi/2}$

$= \left(-\frac{1}{\sqrt{2}} + 1 \right) + \left(1 - \frac{1}{\sqrt{2}} \right)$

$= 2 - \sqrt{2}$

> **COMMENT**
>
> Since the graph is symmetrical about $x = \frac{\pi}{4}$, the area could instead be calculated as $2 \int_0^{\frac{\pi}{4}} \sin x \, dx$.

7 Intersections where

$x(x-4)^2 = x^2 - 7x + 15$

From GDC: $x = 1, 3, 5$

Area enclosed $= \int_1^5 \left| x(x-4)^2 - x^2 + 7x - 15 \right| dx$

$= 8$ (from GDC)

> **COMMENT**
>
> Clearly the answer could be obtained by integrating term by term in the usual way; however, since the question is evidently intended to be answered using a GDC, this is not necessary.

8 Intersections when

$x^2 = mx$

$x(x - m) = 0$

$x = 0$ or m

Enclosed area $= \dfrac{32}{3}$

$\therefore \displaystyle\int_0^m mx - x^2 \, dx = \dfrac{32}{3}$

$\left[\dfrac{mx^2}{2} - \dfrac{x^3}{3} \right]_0^m = \dfrac{32}{3}$

$\dfrac{m^3}{2} - \dfrac{m^3}{3} = \dfrac{32}{3}$

$\dfrac{m^3}{6} = \dfrac{32}{3}$

$m^3 = 64$

$m = 4$

Mixed examination practice 13

Short questions

1 $f(x) = -\cos x + c$

$f\left(\dfrac{\pi}{3}\right) = 0$

$\Rightarrow -\dfrac{1}{2} + c = 0$

$\Rightarrow c = \dfrac{1}{2}$

$\therefore f(x) = \dfrac{1}{2} - \cos x$

2 Intersections with the x-axis when

$k^2 - x^2 = 0$

$x = \pm k$

$\therefore$ Area $= \displaystyle\int_{-k}^{k} k^2 - x^2 \, dx$

$= \left[k^2 x - \dfrac{x^3}{3} \right]_{-k}^{k}$

$= \left(k^3 - \dfrac{k^3}{3} \right) - \left(-k^3 + \dfrac{k^3}{3} \right)$

$= \dfrac{4k^3}{3}$

3 $\displaystyle\int \dfrac{1 + x^2 \sqrt{x}}{x} \, dx = \int x^{-1} + x^{\frac{3}{2}} \, dx$

$= \ln|x| + \dfrac{2x^{\frac{5}{2}}}{5} + c$

4 a $\displaystyle\int_0^a x^3 - x \, dx = 0$

$\left[\dfrac{x^4}{4} - \dfrac{x^2}{2} \right]_0^a = 0$

$\dfrac{a^2}{4}\left(a^2 - 2 \right) = 0$

$a = 0$ or $\pm\sqrt{2}$

Since $a > 0$, $a = \sqrt{2}$

b Curve intersects the x-axis where

$x^3 - x = 0$

$x(x^2 - 1) = 0$

$x = 0$ or ± 1

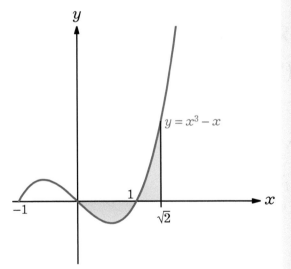

Figure 13MS.4

Total area enclosed = area above x-axis + area below x-axis

Since the integral from 0 to a equals zero (defined in (a)), the area above the x-axis must equal the area below.

∴ Total area = $\left|\int_0^1 x^3 - x\,dx\right| + \left|\int_1^{\sqrt{2}} x^3 - x\,dx\right|$

$= -2\int_0^1 x^3 - x\,dx$

$= -2\left[\dfrac{x^4}{4} - \dfrac{x^2}{2}\right]_0^1$

$= \dfrac{1}{2}$

5 a i Upper right vertex is (a, a^n)

∴ Area $= a \times a^n = a^{n+1}$

ii $B = \int_a^b x^n\,dx$

$= \left[\dfrac{x^{n+1}}{n+1}\right]_a^b$

$= \dfrac{1}{n+1}(b^{n+1} - a^{n+1})$

b The question has been amended to state that the red area is three times as much as the blue area.

Let the red area be A; then

$A = 3B$

$\Rightarrow b \times b^n - a^{n+1} - B = 3B$

$\Rightarrow 4B = b^{n+1} - a^{n+1}$

i.e. $\dfrac{4}{n+1}(b^{n+1} - a^{n+1}) = b^{n+1} - a^{n+1}$

∴ $n + 1 = 4$

$n = 3$

6 The graphs intersect when

$\sin x = 1 - \sin x$

$2\sin x = 1$

$\sin x = \dfrac{1}{2}$

$x = \dfrac{\pi}{6}, \dfrac{5\pi}{6}$ (for $0 < x < \pi$)

Difference function is

$y_1 - y_2 = \sin x - (1 - \sin x) = 2\sin x - 1$

Enclosed area $= \int_{\pi/6}^{5\pi/6} 2\sin x - 1\,dx$

$= [-2\cos x - x]_{\pi/6}^{5\pi/6}$

$= \left(\sqrt{3} - \dfrac{5\pi}{6}\right) - \left(-\sqrt{3} - \dfrac{\pi}{6}\right)$

$= 2\sqrt{3} - \dfrac{2\pi}{3}$

7 a $f''(3) = 24 > 0 \Rightarrow (3, 19)$ is a local minimum

b $f'(x) = \int 6x + 6\,dx$

$= 3x^2 + 6x + c$

Stationary point at $x = 3$

$\Rightarrow f'(3) = 0$

$\Rightarrow 27 + 18 + c = 0$

$\Rightarrow c = -45$

∴ $f'(x) = 3x^2 + 6x - 45$

$f(x) = \int 3x^2 + 6x - 45\,dx$

$= x^3 + 3x^2 - 45x + d$

$f(3) = 19$

$\Rightarrow 27 + 27 - 135 + d = 19$

$\Rightarrow d = 100$

∴ $f(x) = x^3 + 3x^2 - 45x + 100$

Long questions

1 a The tangent to the graph at $x = 1$ has gradient 3, so $f'(1) = 3$

∴ $e^1 + c = 3$

$\Rightarrow c = 3 - e$

b From the tangent equation,

$y(1) = 3 \times 1 + 2 = 5$

∴ $f(1) = 5$

142 Mixed examination practice 13

c $f'(x) = e^x + 3 - e$

$f(x) = \int e^x + 3 - e \, dx$
$= e^x + (3-e)x + d$

$f(1) = 5$
$\Rightarrow e + (3-e) + d = 5$
$\Rightarrow d = 2$

$\therefore f(x) = e^x + (3-e)x + 2$

d $f(x) > 0$ for all $x > 0$

$\therefore$ Area under graph $= \int_0^1 f(x) \, dx$

$= \int_0^1 e^x + (3-e)x + 2 \, dx$

$= \left[e^x + \frac{(3-e)}{2} x^2 + 2x \right]_0^1$

$= \left(e + \frac{3-e}{2} + 2 \right) - 1$

$= \frac{e+5}{2}$

2 a $(5a-x)(x+a) = 5a(x+a) - x(x+a)$
$= 5ax + 5a^2 - x^2 - ax$
$= 5a^2 + 4ax - x^2$

b Intersection when

$5a^2 + 4ax - x^2 = x^2 - a^2$
$(5a-x)(x+a) = (x+a)(x-a)$
$(x+a)[(x-a) - (5a-x)] = 0$
$(x+a)(2x - 6a) = 0$
$(x+a)(x - 3a) = 0$
$(-a, 0)$ and $(3a, 8a^2)$

The coordinates of the points of intersection are $(-a, 0)$ and $(3a, 8a^2)$.

c Difference function is
$y_1 - y_2 = 6a^2 + 4ax - 2x^2$

Area enclosed $= \int_{-a}^{3a} 6a^2 + 4ax - 2x^2 \, dx$

$= \left[6a^2 x + 2ax^2 - \frac{2}{3} x^3 \right]_{-a}^{3a}$

$= (18a^3 + 18a^3 - 18a^3) - \left(-6a^3 + 2a^3 + \frac{2}{3} a^3 \right)$

$= \frac{64}{3} a^3$

d

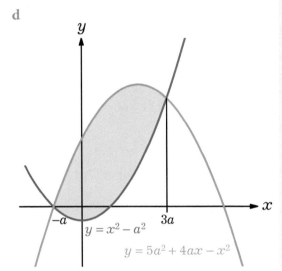

Figure 13ML.2

Area below axis $= \left| \int_{-a}^{a} x^2 - a^2 \, dx \right|$

$= \left| \left[\frac{x^3}{3} - a^2 x \right]_{-a}^{a} \right|$

$= \left| \left(\frac{a^3}{3} - a^3 \right) - \left(-\frac{a^3}{3} + a^3 \right) \right|$

$= \frac{4}{3} a^3$

$\therefore$ Area above axis $= \frac{64}{3} a^3 - \frac{4}{3} a^3 = \frac{60}{3} a^3$

Fraction of the enclosed area which lies above the x-axis is $\dfrac{\left(\dfrac{60a^3}{4} \right)}{\left(\dfrac{64a^3}{4} \right)} = \dfrac{60}{64} = \dfrac{15}{16}$

14 Further differentiation

Exercise 14A

5 $y = (3x+5)^2$

$\dfrac{dy}{dx} = 2(3x+5) \times 3 = 6(3x+5)$

$\therefore \dfrac{dy}{dx}(2) = 6 \times 11 = 66$

$y(2) = (11)^2 = 121$

Equation of tangent at (2, 11) is

$y - y_1 = m(x - x_1)$

$y - 121 = 66(x - 2)$

$y = 66x - 11$

6 $y = (4x^2 + 1)^{-\frac{1}{2}}$

$\dfrac{dy}{dx} = -\dfrac{1}{2}(4x^2 + 1)^{-\frac{3}{2}} \times 8x$

$= -\dfrac{4x}{\sqrt{(4x^2+1)^3}}$

$\therefore \dfrac{dy}{dx}(\sqrt{2}) = -\dfrac{4\sqrt{2}}{\sqrt{9^3}} = -\dfrac{4\sqrt{2}}{27}$

$y(\sqrt{2}) = \dfrac{1}{\sqrt{4 \times 2 + 1}} = \dfrac{1}{3}$

Normal at $\left(\sqrt{2}, \dfrac{1}{3}\right)$ has gradient

$\dfrac{27}{4\sqrt{2}} = \dfrac{27\sqrt{2}}{8}$ and is given by

$y - y_1 = m(x - x_1)$

$y - \dfrac{1}{3} = \dfrac{27\sqrt{2}}{8}(x - \sqrt{2})$

$24y - 8 = 81\sqrt{2}x - 162$

$24y = 81\sqrt{2}x - 154$

$y = \dfrac{27\sqrt{2}}{8}x - \dfrac{77}{12}$

7 $y = e^{\sin x}$

$\Rightarrow \dfrac{dy}{dx} = e^{\sin x} \cos x$

Stationary points where $\dfrac{dy}{dx} = 0$:

$e^{\sin x} \cos x = 0$

$\Rightarrow \cos x = 0$ (since $e^x \neq 0$)

$\therefore x = \dfrac{\pi}{2}, \dfrac{3\pi}{2}$ (for $x \in [0, 2\pi]$)

Coordinates of the stationary points are

$\left(\dfrac{\pi}{2}, e\right)$ and $\left(\dfrac{3\pi}{2}, e^{-1}\right)$

8 $f(x) = \ln(x^2 - 35)$

$f'(x) = (2x) \times \dfrac{1}{x^2 - 35}$

$f'(x) = 1$

$\dfrac{2x}{x^2 - 35} = 1$

$x^2 - 2x - 35 = 0$

$(x+5)(x-7) = 0$

$x = -5$ or 7

The domain of $f(x)$ is $x > \sqrt{35}$, so the only solution is $x = 7$.

9 a $h(-1) = e^{-1} + e^2$

$h(2) = e^2 + e^{-4}$

Since $e^{-1} > e^{-4}$, the post at $x = -1$ is taller.

b $\dfrac{dh}{dx} = e^x - 2e^{-2x}$

Stationary point occurs where $\dfrac{dh}{dx} = 0$:

$e^x - 2e^{-2x} = 0$

$e^x = 2e^{-2x}$

$e^{3x} = 2$

$3x = \ln 2$

$x = \dfrac{1}{3}\ln 2$

Classifying the stationary point:

$\dfrac{d^2h}{dx^2} = e^x + 4e^{-2x}$

$\dfrac{d^2h}{dx^2}\left(\dfrac{1}{3}\ln 2\right) = e^{\frac{1}{3}\ln 2} + 4e^{-\frac{2}{3}\ln 2} > 0$

$\Rightarrow$ local minimum

Therefore the minimum height occurs at $x = \dfrac{1}{3}\ln 2$

c Minimum height is

$h\left(\dfrac{1}{3}\ln 2\right) = h\left(\ln 2^{\frac{1}{3}}\right)$

$= 2^{\frac{1}{3}} + 2^{-\frac{2}{3}}$

$= 2^{\frac{1}{3}}\left(1 + \dfrac{1}{2}\right)$

$= \dfrac{3}{2} \times 2^{\frac{1}{3}}$

$= \dfrac{3\sqrt[3]{2}}{2}$

Exercise 14B

4 Let $u = (3x^2 - x + 2)$, $v = e^{2x}$

$\dfrac{dy}{dx} = v\dfrac{du}{dx} + u\dfrac{dv}{dx}$

$= e^{2x}(6x - 1) + (3x^2 - x + 2) \times 2e^{2x}$

$= (6x^2 + 4x + 3)e^{2x}$

5 Let $u = x$, $v = e^x$

$\dfrac{dy}{dx} = v\dfrac{du}{dx} + u\dfrac{dv}{dx}$

$= e^x + xe^x$

$\dfrac{dy}{dx}(1) = 2e$

$y(1) = e$

Equation of the tangent is

$y - y_1 = m(x - x_1)$

$y - e = 2e(x - 1)$

$y = 2ex - e$

6 Let $u = x^2$, $v = e^{3x}$

$f'(x) = v\dfrac{du}{dx} + u\dfrac{dv}{dx}$

$= e^{3x}(2x) + x^2 \times 3e^{3x}$

$= (2x + 3x^2)e^{3x}$

Let $p = 2x + 3x^2$, $q = e^{3x}$

$f''(x) = q\dfrac{dp}{dx} + p\dfrac{dq}{dx}$

$= e^{3x}(2 + 6x) + (2x + 3x^2) \times 3e^{3x}$

$= (9x^2 + 12x + 2)e^{3x}$

7 Let $u = (2x+1)^5$, $v = e^{-2x}$

$$\frac{dy}{dx} = v\frac{du}{dx} + u\frac{dv}{dx}$$
$$= e^{-2x} \times 5(2x+1)^4 \times 2 + (2x+1)^5(-2e^{-2x})$$
$$= e^{-2x}(2x+1)^4(10-4x-2)$$
$$= e^{-2x}(8-4x)(2x+1)^4$$

Stationary points where $\frac{dy}{dx} = 0$:

$e^{-2x}(8-4x)(2x+1)^4 = 0$
$(8-4x)(2x+1)^4 = 0$
$x = 2$ or $-\frac{1}{2}$

8 Let $u = x$, $v = (x+1)^{\frac{1}{2}}$

$$\frac{dy}{dx} = v\frac{du}{dx} + u\frac{dv}{dx}$$
$$= (x+1)^{\frac{1}{2}} \times 1 + x \times \frac{1}{2}(x+1)^{-\frac{1}{2}}$$
$$= \frac{1}{\sqrt{x+1}}\left(x+1+\frac{x}{2}\right)$$
$$= \frac{1}{2\sqrt{x+1}}(3x+2)$$

Stationary points where $\frac{dy}{dx} = 0$:

$\frac{1}{2\sqrt{x+1}}(3x+2) = 0$
$3x+2 = 0$
$x = -\frac{2}{3}$

$y\left(-\frac{2}{3}\right) = \left(-\frac{2}{3}\right)\sqrt{\frac{1}{3}} = -\frac{2\sqrt{3}}{9}$

∴ coordinates of the stationary point are
$\left(-\frac{2}{3}, -\frac{2\sqrt{3}}{9}\right)$

9 Let $u = (3x+1)^5$, $v = (3-x)^3$

$$\frac{dy}{dx} = v\frac{du}{dx} + u\frac{dv}{dx}$$
$$= (3-x)^3 \times 5(3x+1)^4 \times 3 + (3x+1)^5$$
$$\quad \times 3(3-x)^2 \times (-1)$$
$$= (3-x)^2(3x+1)^4(15(3-x) - 3(3x+1))$$
$$= (3-x)^2(3x+1)^4(42-24x)$$
$$= 6(3-x)^2(3x+1)^4(7-4x)$$

Stationary points where $\frac{dy}{dx} = 0$:

$6(3-x)^2(3x+1)^4(7-4x) = 0$

$x = 3, -\frac{1}{3}, \frac{7}{4}$

10 a Let $u = x$, $v = \sin 2x$

$$\frac{dy}{dx} = v\frac{du}{dx} + u\frac{dv}{dx}$$
$$= \sin 2x \times 1 + x \times 2\cos 2x$$
$$= \sin 2x + 2x\cos 2x$$

$$\frac{d^2y}{dx^2} = 2\cos 2x + 2\cos 2x - 4x\sin 2x$$
$$= 4\cos 2x - 4x\sin 2x$$

Inflexion points where $\frac{d^2y}{dx^2} = 0$:

$4\cos 2x - 4x\sin 2x = 0$
$\cos 2x = x\sin 2x$

b $\cos 2x = x\sin 2x$

$\Rightarrow \tan 2x = \frac{1}{x}$

From GDC: $x = 0.538, 1.82, 3.29, 4.81$

Inflexion points are $(0.538, 0.474)$, $(1.82, -0.877)$, $(3.29, 0.957)$, $(4.81, -0.979)$

146 Topic 14B Differentiating products using the product rule

COMMENT

To get 3SF accuracy on the y-coordinates, ensure that when inserting the x-coordinate into the function you use either the full x-value obtained from your GDC (saved in its memory) or several significant figures beyond the three written down in your answer.

11 Let $w = xe^x$, $u = x$, $v = e^x$

By the product rule,

$$\frac{dw}{dx} = v\frac{du}{dx} + u\frac{dv}{dx}$$

$$= e^x + xe^x$$

Then, by the chain rule, since $y = \sin(w)$:

$$\frac{dy}{dx} = \frac{dw}{dx} \times \cos(w)$$

$$= (e^x + xe^x)\cos(xe^x)$$

$$= e^x(1+x)\cos(xe^x)$$

12 a Let $u = x$, $v = \ln x$

$$f'(x) = v\frac{du}{dx} + u\frac{dv}{dx}$$

$$= \ln x \times 1 + x \times \frac{1}{x}$$

$$= \ln x + 1$$

b $\int \ln x \, dx = \int (\ln x + 1) - 1 \, dx$

$$= \int \ln x + 1 \, dx - \int 1 \, dx$$

$$= x \ln x - x + c$$

13 Let $u = e^{-x}$, $v = \cos x$

$$\frac{dy}{dx} = v\frac{du}{dx} + u\frac{dv}{dx}$$

$$= -e^{-x}\cos x - e^{-x}\sin x$$

$$= -e^{-x}(\sin x + \cos x)$$

Stationary points where $\frac{dy}{dx} = 0$:

$-e^{-x}(\sin x + \cos x) = 0$

$\sin x + \cos x = 0$

$\sin x = -\cos x$

$\tan x = -1$

$\therefore x = \frac{3\pi}{4}$ (for $0 \leq x \leq \pi$)

$$y\left(\frac{3\pi}{4}\right) = e^{-\frac{3\pi}{4}} \cos\left(\frac{3\pi}{4}\right) = -\frac{\sqrt{2}}{2} e^{-\frac{3\pi}{4}}$$

∴ coordinates of the stationary point are

$$\left(\frac{3\pi}{4}, -\frac{\sqrt{2}}{2} e^{-\frac{3\pi}{4}}\right)$$

14 Let $u = x^2$, $v = (1+x)^{\frac{1}{2}}$

$$f'(x) = v\frac{du}{dx} + u\frac{dv}{dx}$$

$$= (1+x)^{\frac{1}{2}} \times 2x + x^2 \times \frac{1}{2}(1+x)^{-\frac{1}{2}}$$

$$= \frac{x}{2(1+x)^{\frac{1}{2}}}(4(1+x) + x)$$

$$= \frac{x}{2\sqrt{1+x}}(4 + 5x)$$

$\therefore a = 4$, $b = 5$

15 a $y = x^x$

$$= (e^{\ln x})^x$$

$$= e^{x \ln x}$$

b Let $u = x$, $v = \ln x$

By the product rule,

$$\frac{d}{dx}(x \ln x) = v\frac{du}{dx} + u\frac{dv}{dx}$$

$$= \ln x \times 1 + x \times \frac{1}{x}$$

$$= \ln x + 1$$

$\therefore \frac{dy}{dx} = (\ln x + 1)e^{x \ln x} = (\ln x + 1)x^x$

14 Further differentiation 147

c Stationary points where $\frac{dy}{dx} = 0$:

$(\ln x + 1)x^x = 0$

$\ln x + 1 = 0$ (as $x^x \neq 0$)

$\ln x = -1$

$x = e^{-1}$

∴ coordinates of the stationary point are $\left(e^{-1}, e^{-e^{-1}}\right)$

Exercise 14C

2 Let $u = x^2$, $v = 2x - 1$

$\frac{dy}{dx} = \frac{v\frac{du}{dx} - u\frac{dv}{dx}}{v^2}$

$= \frac{2x(2x-1) - 2x^2}{(2x-1)^2}$

$= \frac{2x^2 - 2x}{(2x-1)^2}$

$= \frac{2x(x-1)}{(2x-1)^2}$

Stationary points where $\frac{dy}{dx} = 0$:

$\frac{2x(x-1)}{(2x-1)^2} = 0$

$2x(x-1) = 0$

$x = 0$ or 1

$y(0) = 0$, $y(1) = \frac{1^2}{2(1)-1} = 1$

∴ coordinates of the stationary points are $(0, 0)$ and $(1, 1)$

3 $y\left(\frac{\pi}{2}\right) = \frac{\sin\left(\frac{\pi}{2}\right)}{\frac{\pi}{2}} = \frac{2}{\pi}$

Let $u = \sin x$, $v = x$

$\frac{dy}{dx} = \frac{v\frac{du}{dx} - u\frac{dv}{dx}}{v^2}$

$= \frac{x\cos x - \sin x}{x^2}$

$\frac{dy}{dx}\left(\frac{\pi}{2}\right) = \frac{-1}{\left(\frac{\pi}{2}\right)^2} = -\frac{4}{\pi^2}$

Normal at $\left(\frac{\pi}{2}, \frac{2}{\pi}\right)$ has gradient $\frac{\pi^2}{4}$

Equation of the normal is

$y - y_1 = m(x - x_1)$

$y - \frac{2}{\pi} = \frac{\pi^2}{4}\left(x - \frac{\pi}{2}\right)$

$y = \frac{\pi^2}{4}x - \frac{\pi^3}{8} + \frac{2}{\pi}$

$y = \frac{\pi^2}{4}x + \frac{16 - \pi^4}{8\pi}$

4 Let $u = x - a$, $v = x + 2$

$\frac{dy}{dx} = \frac{v\frac{du}{dx} - u\frac{dv}{dx}}{v^2}$

$= \frac{(x+2) - (x-a)}{(x+2)^2} = \frac{a+2}{(x+2)^2}$

$\frac{dy}{dx}(a) = \frac{a+2}{(a+2)^2} = \frac{1}{a+2}$

Require $\frac{dy}{dx}(a) = 1$

∴ $\frac{1}{a+2} = 1$

$a + 2 = 1$

$a = -1$

5 Let $u = \ln x$, $v = x$

$$\frac{dy}{dx} = \frac{v\frac{du}{dx} - u\frac{dv}{dx}}{v^2}$$

$$= \frac{\frac{x}{x} - \ln x}{x^2} = \frac{1 - \ln x}{x^2}$$

Stationary points where $\frac{dy}{dx} = 0$:

$$\frac{1 - \ln x}{x^2} = 0$$

$\ln x = 1$

$x = e$

Let $p = 1 - \ln x$, $q = x^2$

$$\frac{d^2 y}{dx^2} = \frac{q\frac{dp}{dx} - p\frac{dq}{dx}}{q^2}$$

$$= \frac{x^2\left(-\frac{1}{x}\right) - 2x(1 - \ln x)}{x^4}$$

$$= \frac{-x - 2x + 2x\ln x}{x^4}$$

$$= \frac{2\ln x - 3}{x^3}$$

$\frac{d^2 y}{dx^2}(e) = -\frac{1}{e^3} < 0 \Rightarrow$ local maximum

$\therefore$ stationary point at $\left(e, \frac{1}{e}\right)$ is a local maximum

6 Let $u = x^2$, $v = 1 - x$

$$f'(x) = \frac{v\frac{du}{dx} - u\frac{dv}{dx}}{v^2}$$

$$= \frac{2x(1-x) + x^2}{(1-x)^2}$$

$$= \frac{2x - x^2}{(1-x)^2}$$

For the function to be increasing, require $\frac{dy}{dx} > 0$

$$\therefore \frac{2x - x^2}{(1-x)^2} > 0$$

$2x - x^2 > 0$

$x(2 - x) > 0$

$0 < x < 2$

Checking for validity of the function:
$x = 1$ is not in the domain.

So $f(x)$ is increasing for $x \in {]}0,1{[} \cup {]}1,2{[}$.

7 Let $u = x^2$, $v = (x+1)^{\frac{1}{2}}$

$$\frac{dy}{dx} = \frac{v\frac{du}{dx} - u\frac{dv}{dx}}{v^2}$$

$$= \frac{2x(1+x)^{\frac{1}{2}} - \frac{x^2}{2}(x+1)^{-\frac{1}{2}}}{(x+1)}$$

$$= \frac{4x(1+x) - x^2}{2(x+1)^{\frac{3}{2}}}$$

$$= \frac{4x + 3x^2}{2(x+1)^{\frac{3}{2}}}$$

$$= \frac{x(3x + 4)}{2(x+1)^{\frac{3}{2}}}$$

$\therefore a = 3$, $b = 4$, $p = \frac{3}{2}$

8 $f(x)$ has a local maximum at $x = a$

$\Rightarrow f'(a) = 0$ and, for small $\delta > 0$,
$f'(a - \delta) > 0$ and $f'(a + \delta) < 0$

Let $y = \frac{1}{f(x)}$

By the quotient rule or chain rule,

$$\frac{dy}{dx} = -\frac{f'(x)}{(f(x))^2}$$

Then $\dfrac{dy}{dx}(a) = \dfrac{f'(a)}{(f(a))^2} = \dfrac{0}{(f(a))^2} = 0$,

so there is a stationary point at $\left(a, \dfrac{1}{f(a)}\right)$

Also, $\dfrac{dy}{dx}(a-\delta) = -\dfrac{f'(a-\delta)}{(f(a-\delta))^2} < 0$

and $\dfrac{dy}{dx}(a+\delta) = -\dfrac{f'(a+\delta)}{(f(a+\delta))^2} > 0$

Therefore $\left(a, \dfrac{1}{f(a)}\right)$ is a local minimum.

> **COMMENT**
>
> It might seem better to use the second derivative to determine the nature of the stationary point, but there are good reasons not to do this here. It is possible for a local maximum to have a zero second derivative, such as in the curve $y = -x^4$, so any proof would have to include contingencies for this circumstance; and in any case calculating the second derivative of $y = \dfrac{1}{f(x)}$ requires multiple uses of the chain rule, product rule and quotient rule, which is unnecessarily complex.

Exercise 14D

2 a The base of the box is a square with side length $12 - 2x$, and the box sides have height x

$\therefore V = x(12 - 2x)^2$

b Let $u = x$, $v = (12 - 2x)^2$

$\dfrac{dV}{dx} = v\dfrac{du}{dx} + u\dfrac{dv}{dx}$

$= (12 - 2x)^2 - 4x(12 - 2x)$

$= (12 - 2x)(12 - 2x - 4x)$

$= (12 - 2x)(12 - 6x)$

$= 12(6 - x)(2 - x)$

Stationary values for V occur when $\dfrac{dV}{dx} = 0$:

$12(6 - x)(2 - x) = 0$

$\Rightarrow x = 6$ or 2

Clearly $x = 6$ represents a minimum, with zero volume, so $x = 2$ must represent the maximum.

Checking using the second derivative:

$\dfrac{d^2V}{dx^2} = -12(6 - x) - 12(2 - x) = -12(8 - 2x)$

$\dfrac{d^2V}{dx^2}(2) = -48 < 0$, so volume is indeed maximal at $x = 2$.

3 Let each side of the base have length x, and let the height of the box be h.

The surface area is $S = x^2 + 4xh$

The volume is $V = x^2 h$

$V = 32$

$\Rightarrow x^2 h = 32$

$\Rightarrow h = \dfrac{32}{x^2}$

Substituting into the expression for S:

$S = x^2 + 4x\left(\dfrac{32}{x^2}\right)$

$= x^2 + \dfrac{128}{x}$

$\dfrac{dS}{dx} = 2x - \dfrac{128}{x^2}$

Stationary values of S occur when $\dfrac{dS}{dx} = 0$:

$2x - \dfrac{128}{x^2} = 0$

$2x = \dfrac{128}{x^2}$

$x^3 = 64$

$x = 4$

Check that this is a local minimum:

$\dfrac{d^2S}{dx^2} = 2 + \dfrac{256}{x^3}$

$\dfrac{d^2S}{dx^2}(4) = 6 > 0 \Rightarrow$ local minimum

$\therefore S(4) = 4^2 + \dfrac{128}{4} = 16 + 32 = 48 \text{ cm}^2$ is the minimum surface area.

4 Let point A have coordinates $(x, 0)$ where $x > 0$. Then:

Area of rectangle $= 2x(4 - x^2)$

$= 8x - 2x^3$

$\dfrac{d\text{ Area}}{dx} = 8 - 6x^2$

Stationary value when $\dfrac{d\text{ Area}}{dx} = 0$:

$8 - 6x^2 = 0$

$x^2 = \dfrac{4}{3}$

$x = \dfrac{2}{\sqrt{3}}$

It is clear that this is a maximum rather than a minimum, from the context.

Showing this rigorously:

$\dfrac{d^2\text{Area}}{dx^2} = -12x$

$\dfrac{d^2\text{Area}}{dx^2}\left(\dfrac{2}{\sqrt{3}}\right) < 0 \Rightarrow$ local maximum

Hence the x-coordinate of A that gives the maximum possible area is $\dfrac{2}{\sqrt{3}} = \dfrac{2\sqrt{3}}{3}$

5 The question has been amended to state that A is $(x, \sin x)$.

a i By symmetry, the coordinates of B are $(\pi - x, \sin x)$

ii Area $= (\pi - 2x)\sin x$

b $\dfrac{d\text{ Area}}{dx} = (\pi - 2x)\cos x - 2\sin x$

Area has a stationary value when $\dfrac{d\text{ Area}}{dx} = 0$:

$(\pi - 2x)\cos x - 2\sin x = 0$

$2\sin x = (\pi - 2x)\cos x$

$2\tan x = \pi - 2x$

c From GDC: $x = 0.710$

Hence maximum area is

$(\pi - 0.710)\sin(0.710) = 1.12 \text{ (3SF)}$

6 Let the radius be r and the height h.

Volume $V = \pi r^2 h$

Surface area $S = 2\pi r^2 + 2\pi rh$

$\therefore S = 450$

$\Rightarrow 2\pi r^2 + 2\pi rh = 450$

$2\pi r(r + h) = 450$

$\Rightarrow h = \dfrac{450}{2\pi r} - r$

Substituting into the expression for volume:

$V = \pi r^2 \left(\dfrac{450}{2\pi r} - r\right)$

$= 225r - \pi r^3$

$$\therefore \frac{dV}{dr} = 225 - 3\pi r^2$$

Stationary values of V occur when $\frac{dV}{dr} = 0$:

$$225 - 3\pi r^2 = 0$$
$$3\pi r^2 = 225$$
$$\therefore r = \sqrt{\frac{75}{\pi}}$$

$$\frac{d^2V}{dr^2} = -6\pi r$$

$$\frac{d^2V}{dr^2}\left(\sqrt{\frac{75}{\pi}}\right) < 0 \Rightarrow \text{local maximum}$$

So largest possible capacity is

$$V\left(\sqrt{\frac{75}{\pi}}\right) = \sqrt{\frac{75}{\pi}}(225 - 75) = 733 \text{ cm}^3 \text{ (3SF)}$$

7 Let each side of the base have length x, and let the height be h.

Volume $V = x^2 h$

Surface area $S = 2x^2 + 4xh$

$$\therefore S = 450$$
$$\Rightarrow 2x^2 + 4xh = 450$$
$$4x\left(\frac{x}{2} + h\right) = 450$$
$$\Rightarrow h = \frac{450}{4x} - \frac{x}{2}$$

Substituting into the expression for volume:

$$V = x^2\left(\frac{450}{4x} - \frac{x}{2}\right)$$
$$= \frac{1}{2}(225x - x^3)$$

$$\frac{dV}{dx} = \frac{1}{2}(225 - 3x^2)$$

Stationary values of V occur when $\frac{dV}{dx} = 0$:

$$\frac{1}{2}(225 - 3x^2) = 0$$
$$3x^2 = 225$$
$$\therefore x = \sqrt{75}$$

$$\frac{d^2V}{dx^2} = -3x$$

$$\frac{d^2V}{dx^2}(\sqrt{75}) < 0 \Rightarrow \text{local maximum}$$

So largest possible capacity is

$$V(\sqrt{75}) = \frac{\sqrt{75}}{2}(225 - 75)$$
$$= 75\sqrt{75}$$
$$= 650 \text{ cm}^3 \text{ (3SF)}$$

8 $x + y = 6$
$$\Rightarrow y = 6 - x$$

$$S = x^2 + y^2$$
$$= x^2 + (6 - x)^2$$
$$= 2x^2 - 12x + 36$$

$$\frac{dS}{dx} = 4x - 12$$

Stationary values of S when $\frac{dS}{dx} = 0$:

$$4x - 12 = 0$$
$$\Rightarrow x = 3$$

$$\frac{d^2S}{dx^2} = 4 > 0 \Rightarrow \text{local minimum}$$

a $x = 3$, $y = 3$ gives the minimum value of the sum of squares.

b End-point values of 0 and 6 produce the maximum value of the sum of squares.

9 a Curved surface area of the cone is given by $S = \pi r \sqrt{r^2 + h^2}$

Volume $V = 81\pi$

$\Rightarrow \dfrac{1}{3}\pi r^2 h = 81\pi$

$\Rightarrow h = \dfrac{243}{r^2}$

Substituting into the expression for surface area:

$S = \pi r \sqrt{r^2 + \dfrac{243^2}{r^4}}$

$= \dfrac{\pi r}{r^2}\sqrt{r^6 + 243^2}$

$= \dfrac{\pi}{r}\sqrt{r^6 + 243^2}$

b $S = \pi\left(r^4 + 243^2 r^{-2}\right)^{\frac{1}{2}}$

$\Rightarrow \dfrac{dS}{dr} = \dfrac{\pi}{2}\left(4r^3 - 2\times 243^2 r^{-3}\right)\left(r^4 + 243^2 r^{-2}\right)^{-\frac{1}{2}}$

Stationary values of S when $\dfrac{dS}{dr} = 0$:

$\dfrac{\pi\left(4r^3 - 2\times 243^2 r^{-3}\right)}{2\sqrt{r^4 + 243^2 r^{-2}}} = 0$

$4r^3 - 2\times 243^2 r^{-3} = 0$

$r^6 = \dfrac{243^2}{2}$

$r = 5.56\ (3\text{SF})$

$\therefore h = \dfrac{243}{r^2} = 7.86$

This pair of r and h values clearly produces a minimum value for the surface area rather than a maximum, since the surface area can be made arbitrarily large by taking sufficiently small or large values of r.

10 a The triangle has side lengths b, $10 - \dfrac{b}{2}$, $10 - \dfrac{b}{2}$

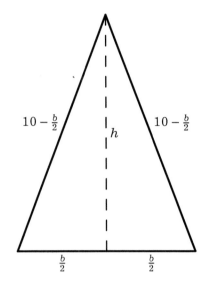

Figure 14D.10

By Pythagoras' Theorem,

$h = \sqrt{\left(10 - \dfrac{b}{2}\right)^2 - \left(\dfrac{b}{2}\right)^2}$

$= \sqrt{100 - 10b}$

Area of triangle is

$A = \dfrac{bh}{2}$

$= \dfrac{b}{2}\sqrt{100 - 10b}$

b $A = \dfrac{b}{2}(100 - 10b)^{\frac{1}{2}}$

$\Rightarrow \dfrac{dA}{db} = \dfrac{1}{2}(100 - 10b)^{\frac{1}{2}} + \dfrac{b}{4}(100 - 10b)^{-\frac{1}{2}} \times (-10)$

$= \dfrac{1}{2\sqrt{100 - 10b}}(100 - 10b - 5b)$

$= \dfrac{1}{2\sqrt{100 - 10b}}(100 - 15b)$

Stationary value of A when $\dfrac{dA}{db} = 0$:

$$\dfrac{1}{2\sqrt{100-10b}}(100-15b) = 0$$

$$100 - 15b = 0$$

$$15b = 100$$

$$b = \dfrac{20}{3}$$

That is, the base length is one-third of the perimeter of the isosceles triangle, so the triangle is equilateral.

This clearly gives a maximum value rather than a minimum, since the area can be made arbitrarily small by taking b small enough or close enough to 10.

11 Let the two numbers be x and y.

$$x^2 + y^2 = a$$

$$\Rightarrow y = \sqrt{a - x^2}$$

The product P is given by

$$P = xy$$

$$= x\sqrt{a - x^2}$$

$$= x(a - x^2)^{\frac{1}{2}}$$

$$\dfrac{dP}{dx} = (a - x^2)^{\frac{1}{2}} - x^2(a - x^2)^{-\frac{1}{2}}$$

$$= \dfrac{1}{\sqrt{a - x^2}}(a - x^2 - x^2)$$

$$= \dfrac{a - 2x^2}{\sqrt{a - x^2}}$$

Stationary values of P when $\dfrac{dP}{dx} = 0$:

$$\dfrac{a - 2x^2}{\sqrt{a - x^2}} = 0$$

$$a - 2x^2 = 0$$

$$2x^2 = a$$

$$\therefore x = \sqrt{\dfrac{a}{2}} \quad \text{(since } x > 0\text{)}$$

$$y^2 = a - x^2$$

$$= a - \dfrac{a}{2} = \dfrac{a}{2}$$

$$\therefore y = \sqrt{\dfrac{a}{2}} \quad \text{(since } y > 0\text{)}$$

Hence $x = y$ for the stationary point of P.

$$\dfrac{d^2P}{dx^2} = \dfrac{-4x(a - x^2)^{\frac{1}{2}} + x(a - 2x^2)(a - x^2)^{-\frac{1}{2}}}{a - x^2}$$

$$= \dfrac{-4x(a - x^2) + x(a - 2x^2)}{(a - x^2)^{\frac{3}{2}}}$$

$$\dfrac{d^2P}{dx^2}\left(\sqrt{\dfrac{a}{2}}\right) = \dfrac{-4\sqrt{\dfrac{a}{2}}\left(\dfrac{a}{2}\right)}{\left(\dfrac{a}{2}\right)^{\frac{3}{2}}} = -4 < 0$$

$\Rightarrow$ local maximum

Hence P is at a maximum when $x = y$.

12

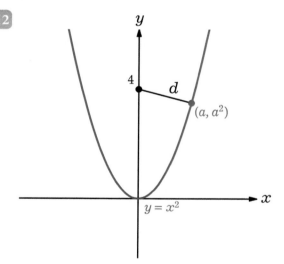

Figure 14D.12

The distance d from point (a, a^2) to $(0, 4)$ is given by

$$d^2 = a^2 + (4 - a^2)^2$$

$$= 16 - 7a^2 + a^4$$

$$\dfrac{d}{da}d^2 = -14a + 4a^3$$

Stationary value of d^2 when $\dfrac{\mathrm{d}}{\mathrm{d}a}d^2 = 0$:

$-14a + 4a^3 = 0$

$2a(2a^2 - 7) = 0$

$2a = 0$ or $a^2 = \dfrac{7}{2}$

$\therefore a = 0$ or $a = \sqrt{\dfrac{7}{2}}$ (since $a \geq 0$)

$\dfrac{\mathrm{d}^2}{\mathrm{d}a^2}d^2 = -14 + 12a^2$

$\dfrac{\mathrm{d}^2}{\mathrm{d}a^2}d^2(0) = -14 < 0 \Rightarrow a = 0$ represents a local maximum for d^2

$\dfrac{\mathrm{d}^2}{\mathrm{d}a^2}d^2\left(\sqrt{\dfrac{7}{2}}\right) = 28 > 0 \Rightarrow a = \sqrt{\dfrac{7}{2}}$ represents a local minimum for d^2

$\therefore$ the point closest to $(0, 4)$ on the curve for $x \geq 0$ is $\left(\sqrt{\dfrac{7}{2}}, \dfrac{7}{2}\right)$

Mixed examination practice 14

Short questions

1 $y = (4 - x^2)^{-1}$

$\Rightarrow \dfrac{\mathrm{d}y}{\mathrm{d}x} = 2x(4 - x^2)^{-2}$

$\dfrac{\mathrm{d}y}{\mathrm{d}x}\left(\dfrac{1}{2}\right) = \dfrac{1}{\left(4 - \dfrac{1}{4}\right)^2} = \dfrac{16}{225}$

2 a $\dfrac{\mathrm{d}y}{\mathrm{d}x} = 5e^{5x}$

b $y = (3x + 2)^{\frac{1}{2}}$

$\Rightarrow \dfrac{\mathrm{d}y}{\mathrm{d}x} = 3 \times \dfrac{1}{2}(3x + 2)^{-\frac{1}{2}} = \dfrac{3}{2\sqrt{3x + 2}}$

c Let $u = e^{5x}$, $v = (3x + 2)^{\frac{1}{2}}$

By the product rule,

$\dfrac{\mathrm{d}y}{\mathrm{d}x} = v\dfrac{\mathrm{d}u}{\mathrm{d}x} + u\dfrac{\mathrm{d}v}{\mathrm{d}x}$

$= 5e^{5x}(3x + 2)^{\frac{1}{2}} + e^{5x} \times \dfrac{3}{2}(3x + 2)^{-\frac{1}{2}}$

$= \dfrac{e^{5x}}{2\sqrt{3x + 2}}(10(3x + 2) + 3)$

$= \dfrac{e^{5x}(30x + 23)}{2\sqrt{3x + 2}}$

3 $y = xe^{-kx}$

$\Rightarrow \dfrac{\mathrm{d}y}{\mathrm{d}x} = e^{-kx} - kxe^{-kx} = (1 - kx)e^{-kx}$

$\dfrac{\mathrm{d}y}{\mathrm{d}x}\left(\dfrac{2}{5}\right) = 0$

$\Rightarrow \left(1 - \dfrac{2k}{5}\right)e^{-\frac{2k}{5}} = 0$

$\Rightarrow 1 - \dfrac{2k}{5} = 0$

$\therefore k = \dfrac{5}{2}$

4 Let $AD = x$. Then $DC = 60 - 2x$ and the area, y, is

$y = x(60 - 2x)$

$= 60x - 2x^2$

$\dfrac{\mathrm{d}y}{\mathrm{d}x} = 60 - 4x$

Maximum area when $\dfrac{\mathrm{d}y}{\mathrm{d}x} = 0$:

$60 - 4x = 0$

$\therefore x = 15$

Hence the maximum area is attained when AD is 15 metres.

This is clearly maximal (not minimal) since the minimal achievable area is zero, when $AD = 0$ or 30. (Alternatively, this can be argued by noting that y is a negative quadratic or by observing that $\dfrac{d^2 y}{dx^2} = -4$, so the only turning point must be a local maximum.)

5 a $f(x) = a(b + e^{-cx})^{-1}$

Using the chain rule:

$$f'(x) = -a(b + e^{-cx})^{-2}(-ce^{-cx}) = ace^{-cx}(b + e^{-cx})^{-2}$$

Using the product rule with $u = ace^{-cx}$, $v = (b + e^{-cx})^{-2}$ and the chain rule again:

$$f''(x) = v\dfrac{du}{dx} + u\dfrac{dv}{dx}$$

$$= -ac^2 e^{-cx}(b + e^{-cx})^{-2} + ace^{-cx}\left(-2(b + e^{-cx})^{-3}\right)(-ce^{-cx})$$

$$= -ac^2 e^{-cx}(b + e^{-cx})^{-2} + 2ac^2 e^{-2cx}(b + e^{-cx})^{-3}$$

$$= \dfrac{ac^2 e^{-cx}}{(b + e^{-cx})^3}\left(-(b + e^{-cx}) + 2e^{-cx}\right)$$

$$= \dfrac{ac^2 e^{-cx}}{(b + e^{-cx})^3}(e^{-cx} - b)$$

b $f''(x) = 0$

$$\dfrac{ac^2 e^{-cx}}{(b + e^{-cx})^3}(e^{-cx} - b) = 0$$

$$e^{-cx} - b = 0$$

$$e^{-cx} = b$$

$$x = -\dfrac{\ln b}{c}$$

$$f\left(-\dfrac{\ln b}{c}\right) = \dfrac{a}{b + e^{\ln b}} = \dfrac{a}{2b}$$

∴ the coordinates of the point where $f''(x) = 0$ are $\left(-\dfrac{\ln b}{c}, \dfrac{a}{2b}\right)$

Long questions

1 a $e^x \sin x = 0$

$\Rightarrow \sin x = 0$

∴ $x = \pi$

b i By the product rule,

$$f'(x) = e^x \sin x + e^x \cos x$$

$$= e^x (\sin x + \cos x)$$

ii B is a stationary point, so $f'(x)=0$ at B.

  c From (b)(i), using the product rule:
     $$f''(x) = e^x(\sin x + \cos x) + e^x(\cos x - \sin x)$$
     $$= 2e^x \cos x$$

  d i Since A is a point of inflexion, $f''(x) = 0$ at A.

    ii $2e^x \cos x = 0$
       $\Rightarrow \cos x = 0$
       $\therefore x = \dfrac{\pi}{2}$

       $f\left(\dfrac{\pi}{2}\right) = e^{\frac{\pi}{2}} \sin\left(\dfrac{\pi}{2}\right) = e^{\frac{\pi}{2}}$

       $\therefore$ the coordinates of A are $\left(\dfrac{\pi}{2}, e^{\frac{\pi}{2}}\right)$

2 a Vertical asymptote where denominator equals zero: $x = \dfrac{1}{2}$

  b By the quotient rule,
     $$\dfrac{dy}{dx} = \dfrac{2x(1-2x) - x^2(-2)}{(1-2x)^2}$$
     $$= \dfrac{-2x^2 + 2x}{(1-2x)^2}$$
     $$= \dfrac{2x(1-x)}{(1-2x)^2}$$

     Stationary points where $\dfrac{dy}{dx} = 0$:
     $$\dfrac{2x(1-x)}{(1-2x)^2} = 0$$
     $2x(1-x) = 0$
     $x = 0$ or 1

     $\therefore$ the stationary points are $(0, 0)$ and $(1, -1)$

  c $\dfrac{dy}{dx} = \dfrac{2x - 2x^2}{(1-2x)^2}$

     By the quotient rule,
     $$\dfrac{d^2y}{dx^2} = \dfrac{(2-4x)(1-2x)^2 + 4(2x-2x^2)(1-2x)}{(1-2x)^4}$$
     $$= \dfrac{(2-4x)(1-2x) + 4(2x-2x^2)}{(1-2x)^3}$$
     $$= \dfrac{2}{(1-2x)^3}$$

     $\dfrac{d^2y}{dx^2}(0) = 2 > 0 \Rightarrow$ local minimum

     $\dfrac{d^2y}{dx^2}(1) = -2 < 0 \Rightarrow$ local maximum

     So $(0, 0)$ is a local minimum and $(1, -1)$ is a local maximum.

  d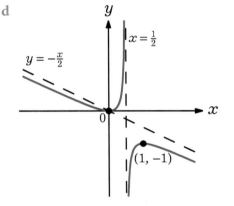

  Figure 14ML.2 Graph of $y = \dfrac{x^2}{1-2x}$

3 a i $2^x = \left(e^{\ln 2}\right)^x = e^{x \ln 2}$

    ii $\dfrac{d}{dx}(2^x) = \dfrac{d}{dx}\left(e^{x \ln 2}\right)$
        $= (\ln 2)\, e^{x \ln 2}$
        $= 2^x \ln 2$

b i By the quotient rule,

$$f'(x) = \frac{2x \times 2^x - x^2 \times 2^x \ln 2}{(2^x)^2}$$

$$= \frac{2x - x^2 \ln 2}{2^x}$$

ii By the quotient rule,

$$f''(x) = \frac{(2 - 2x \ln 2)2^x - (2x - x^2 \ln 2)2^x \ln 2}{(2^x)^2}$$

$$= \frac{2 - 4x \ln 2 + x^2 (\ln 2)^2}{2^x}$$

c i $f'(x) = 0$

$$\frac{2x - x^2 \ln 2}{2^x} = 0$$

$$2x - x^2 \ln 2 = 0$$

$$x(2 - x \ln 2) = 0$$

$$\therefore x = 0 \text{ or } \frac{2}{\ln 2}$$

As $x > 0$, the only solution is $x = \frac{2}{\ln 2}$

ii

$$f''\left(\frac{2}{\ln 2}\right) = \frac{2 - 4\left(\frac{2}{\ln 2}\right)\ln 2 + \left(\frac{4}{(\ln 2)^2}\right)(\ln 2)^2}{2^{\frac{2}{\ln 2}}}$$

$$= \frac{2 - 8 + 4}{2^{\frac{2}{\ln 2}}} < 0$$

so $x = \frac{2}{\ln 2}$ gives a local maximum of $f(x)$

15 Further integration

Exercise 15A

6 Both are correct: their answers differ only in the (unknown) constant.

Marina has $f(x) = \dfrac{1}{3}\ln x + c$

Jack has
$$g(x) = \dfrac{1}{3}\ln 3x + c$$
$$= \dfrac{1}{3}(\ln 3 + \ln x) + c$$
$$= \dfrac{1}{3}\ln x + \dfrac{1}{3}\ln 3 + c$$
$$= \dfrac{1}{3}\ln x + d$$

i.e. the constants are related by
$$d = \dfrac{1}{3}\ln 3 + c$$

7 $\displaystyle\int_{a^2}^{a}(1-x)^{-1}\,dx = 0.4$

$\left[-\ln(1-x)\right]_{a^2}^{a} = 0.4$

$\ln(1-a^2) - \ln(1-a) = 0.4$

$\ln\left(\dfrac{1-a^2}{1-a}\right) = 0.4$

$\ln(1+a) = 0.4$

$1 + a = e^{0.4}$

$a = e^{0.4} - 1$

$= 0.492$ (3SF)

> **COMMENT**
> Note that because $0 < a < 1$, a^2 is the lower limit and a is the upper limit.

Exercise 15B

3 Let $u = \sin x$; then
$$\dfrac{du}{dx} = \cos x$$
$$dx = \dfrac{du}{\cos x}$$

$\displaystyle\int_0^{\pi/2} \cos x\, e^{\sin x}\,dx = \int_{x=0}^{x=\pi/2} \cos x\, e^u \dfrac{du}{\cos x}$

$= \displaystyle\int_{x=0}^{x=\pi/2} e^u\,du$

$= \left[e^u\right]_{x=0}^{x=\pi/2}$

$= \left[e^{\sin x}\right]_0^{\pi/2}$

$= e - 1$

4 Let $u = x^3 + 5$; then
$$\dfrac{du}{dx} = 3x^2$$
$$dx = \dfrac{du}{3x^2}$$

$\displaystyle\int x^2 \sqrt{x^3+5}\,dx = \int x^2 \sqrt{u}\, \dfrac{du}{3x^2}$

$= \dfrac{1}{3}\displaystyle\int u^{\frac{1}{2}}\,du$

$= \dfrac{2}{9} u^{\frac{3}{2}} + c$

$= \dfrac{2}{9}(x^3 + 5)^{\frac{3}{2}} + c$

5 Let $u = e^x + 1$; then

$$\frac{du}{dx} = e^x$$

$$dx = \frac{du}{e^x}$$

$$\int_0^1 \frac{e^x}{\sqrt{e^x+1}} \, dx = \int_{x=0}^{x=1} \frac{e^x}{\sqrt{u}} \frac{du}{e^x}$$

$$= \int_{x=0}^{x=1} u^{-\frac{1}{2}} \, du$$

$$= \left[2u^{\frac{1}{2}} \right]_{x=0}^{x=1}$$

$$= \left[2\sqrt{e^x+1} \right]_0^1$$

$$= 2\sqrt{e+1} - 2\sqrt{2}$$

6 Let $u = x^2 + x - 1$; then

$$\frac{du}{dx} = 2x+1$$

$$dx = \frac{du}{(2x+1)}$$

$$\int_0^2 (2x+1)e^{x^2+x-1} \, dx = \int_{x=0}^{x=2} (2x+1)e^u \frac{du}{(2x+1)}$$

$$= \int_{x=0}^{x=2} e^u \, du$$

$$= \left[e^u \right]_{x=0}^{x=2}$$

$$= \left[e^{x^2+x-1} \right]_0^2$$

$$= e^5 - e^{-1}$$

7 Let $u = x^2 - 1$; then

$$\frac{du}{dx} = 2x$$

$$dx = \frac{du}{2x}$$

$$\int_2^5 \frac{2x}{x^2-1} \, dx = \int_{x=2}^{x=5} \frac{2x}{u} \frac{du}{2x}$$

$$= \int_{x=2}^{x=5} u^{-1} \, du$$

$$= \left[\ln u \right]_{x=2}^{x=5}$$

$$= \left[\ln(x^2-1) \right]_2^5$$

$$= \ln 24 - \ln 3$$

$$= \ln \left(\frac{24}{3} \right)$$

$$= \ln 8$$

8 a $\tan x = \dfrac{\sin x}{\cos x}$

b Let $u = \cos x$; then

$$\frac{du}{dx} = -\sin x$$

$$dx = -\frac{du}{\sin x}$$

$$\int \tan x \, dx = \int \frac{\sin x}{\cos x} \, dx$$

$$= -\int \frac{\sin x}{u} \frac{du}{\sin x}$$

$$= -\int \frac{1}{u} \, du$$

$$= -\ln u + c$$

$$= -\ln(\cos x) + c$$

9 Let $u = x^2 - 3x + 3$; then

$$\frac{du}{dx} = 2x - 3$$

$$dx = \frac{du}{(2x-3)}$$

$$\int_1^3 \frac{(2x-3)\sqrt{x^2-3x+3}}{x^2-3x+3} \, dx = \int_{x=1}^{x=3} \frac{(2x-3)\sqrt{u}}{u} \cdot \frac{du}{(2x-3)}$$

$$= \int_{x=1}^{x=3} u^{-\frac{1}{2}} \, du$$

$$= \left[2u^{\frac{1}{2}} \right]_{x=1}^{x=3}$$

$$= \left[2\sqrt{x^2 - 3x + 3} \right]_1^3$$

$$= 2\sqrt{3} - 2$$

10 All are correct: as in Exercise 15A question 6, the difference lies in the unknown constant.

$$A(x) = \frac{1}{2}\sin^2 x + c$$

$$B(x) = -\frac{1}{2}\cos^2 x + c$$

$$= -\frac{1}{2}(1 - \sin^2 x) + c$$

$$= \frac{1}{2}\sin^2 x + c - \frac{1}{2}$$

$$= \frac{1}{2}\sin^2 x + d$$

which is the same as $A(x)$ with $d = c - \frac{1}{2}$

$$C(x) = -\frac{1}{4}\cos 2x + c$$

$$= -\frac{1}{4}(1 - 2\sin^2 x) + c$$

$$= \frac{1}{2}\sin^2 x + c - \frac{1}{4}$$

$$= \frac{1}{2}\sin^2 x + k$$

which is the same as $A(x)$ with $k = c - \frac{1}{4}$

Exercise 15C

4 i $v = \int a\, dt$

$= at + c$

But $v(0) = u$

$\therefore u = 0 + c$

$\Rightarrow c = u$

$\therefore v = u + at$

ii $s = \int v\, dt$

$= \int u + at\, dt$

$= ut + \dfrac{at^2}{2} + k$

But $s(0) = 0$ (by definition of displacement s being the displacement from the initial position at $t = 0$)

$\therefore 0 = 0 + 0 + k$

$\Rightarrow k = 0$

$\therefore s = ut + \dfrac{1}{2}at^2$

5 a $s(6) = \int_0^6 v\, dt$

$= \int_0^6 5\cos\left(\dfrac{t}{3}\right) dt$

$= \left[15\sin\left(\dfrac{t}{3}\right)\right]_0^6$

$= 15\sin(2)$

$= 13.6$ m (3SF)

b $x(6) = \int_0^6 |v|\, dt$

In [0, 6], $v = 0$ when

$5\cos\left(\dfrac{t}{3}\right) = 0$

$t = \dfrac{3\pi}{2}$

$\therefore x(6) = \int_0^{3\pi/2} v\, dt - \int_{3\pi/2}^6 v\, dt$

$= s(6) - 2\int_{3\pi/2}^6 5\cos\left(\dfrac{t}{3}\right) dt$

$= s(6) - 2\left[15\sin\left(\dfrac{t}{3}\right)\right]_{3\pi/2}^6$

$= 13.6 - 30\left(\sin 2 - \sin\dfrac{\pi}{2}\right)$

$= 16.4$ m (3SF)

> **COMMENT**
>
> Notice the importance of finding where $v = 0$; at such points the velocity graph crosses the t-axis and so the integration needs to be separated at these points (as the 'area' will be negative below the t-axis).

6 a $v = 12 - 9.8t$

$s = \int v\, dt$

$= \int 12 - 9.8t\, dt$

$= 12t - 4.9t^2 + c$

When $t = 0$, $s = 0$ (displacement is measured relative to starting position)

$\therefore c = 0$

$s = 12t - 4.9t^2$

At $t = 2$, $s = 24 - 4.9 \times 4 = 4.4$ m

b Distance travelled in the first 2 seconds is

$x(2) = \int_0^2 |v|\, dt$

In [0, 2], $v = 0$ when

$12 - 9.8t = 0$

$t = \dfrac{12}{9.8} = \dfrac{60}{49}$

so velocity graph goes negative (below the t-axis) for $t > \dfrac{60}{49}$

$\therefore x(2) = \int_0^{60/49} v\,dt - \int_{60/49}^2 v\,dt$

$= \int_0^{60/49} 12 - 9.8t\,dt - \int_{60/49}^2 12 - 9.8t\,dt$

$= 10.3\,\text{m}\,(3\text{SF, from GDC})$

7 a By the quotient rule,

$a = \dfrac{dv}{dt}$

$= \dfrac{(t^2+1) \times 1 - t \times 2t}{(t^2+1)^2}$

$= \dfrac{1-t^2}{(t^2+1)^2}$

b $s(5) = \int_0^5 v\,dt$

$= \int_0^5 \dfrac{t}{t^2+1}\,dt$

Let $w = t^2 + 1$; then

$\dfrac{dw}{dt} = 2t \Rightarrow dt = \dfrac{dw}{2t}$

$\therefore s(5) = \int_{t=0}^{t=5} \dfrac{t}{w} \dfrac{dw}{2t}$

$= \dfrac{1}{2} \int_{t=0}^{t=5} \dfrac{1}{w}\,dw$

$= \dfrac{1}{2}[\ln w]_{t=0}^{t=5}$

$= \dfrac{1}{2}\left[\ln(t^2+1)\right]_0^5$

$= \dfrac{1}{2}(\ln 26 - \ln 1)$

$= \ln\sqrt{26}$

$= 1.63\,(3\text{SF})$

8 $v = \dfrac{ds}{dt} = -t^2 + 3t + 4$

$\Rightarrow \dfrac{dv}{dt} = -2t + 3$

Stationary value of v when $\dfrac{dv}{dt} = 0$:

$-2t + 3 = 0$

$\Rightarrow t = \dfrac{3}{2}$

$v\left(\dfrac{3}{2}\right) = -\left(\dfrac{3}{2}\right)^2 + 3\left(\dfrac{3}{2}\right) + 4 = \dfrac{25}{4}$

Check values at end points:

$v(0) = 4 < v\left(\dfrac{3}{2}\right)$

$v(5) = -6 < v\left(\dfrac{3}{2}\right)$

$\therefore$ maximum velocity is $v(1.5) = 6.25$

9 a $v(0) = 0$

$v(4) = \dfrac{1}{2}(16) - 8 = 0$

$\therefore$ car has zero velocity (is stationary) at $t = 0$ and 4 seconds.

b $v = \dfrac{1}{2}t(t-4)$

v is negative for $0 < t < 4$, so the car is reversing.

v is positive for $t > 4$, so the car is moving forwards.

$\therefore$ the car reverses for 4 seconds, stops at $t = 4$ seconds, and then moves forwards.

c $d(4) = \int_0^4 |v|\,dt$

$= -\int_0^4 \dfrac{t^2}{2} - 2t\,dt$

$= -\left[\dfrac{t^3}{6} - t^2\right]_0^4$

$= 16 - \dfrac{64}{6}$

$= \dfrac{16}{3}$

$= 5.33\,\text{m}\,(3\text{SF})$

d $d(8) = \int_0^8 |v|\, dt$

$= \int_0^4 |v|\, dt + \int_4^8 |v|\, dt$

$= \dfrac{16}{3} + \int_4^8 \dfrac{t^2}{2} - 2t\, dt$

$= \dfrac{16}{3} + \left[\dfrac{t^3}{6} - t^2\right]_4^8$

$= \dfrac{16}{3} + \dfrac{512}{6} - 64 - \dfrac{64}{6} + 16$

$= 32$ m

e $s(8) = \int_0^8 v\, dt$

$= \left[\dfrac{t^3}{6} - t^2\right]_0^8$

$= \dfrac{512}{6} - 64$

$= \dfrac{64}{3}$

$= 21.3$ m

f $a = \dfrac{dv}{dt} = t - 2$

$a = 5 \Rightarrow t = 7$ seconds

g Stationary values of v when $\dfrac{dv}{dt} = 0$:

$t - 2 = 0$

$\therefore t = 2$

$v(2) = -2$ is clearly a minimum velocity (car in reverse)

Check values at end points:

$v(0) = 0$

$v(8) = 16$

So maximum speed is $16\,\text{m s}^{-1}$, at $t = 8$ seconds.

Exercise 15D

3 $V = \pi \int_1^{2e} y^2\, dx$

$= \pi \int_1^{2e} (\ln x)^2\, dx$

$= 19.0$ (from GDC)

4 $V = \pi \int_0^{\pi/2} y^2\, dx$

$= \pi \int_0^{\pi/2} \sin x\, dx$

$= \pi [-\cos x]_0^{\pi/2}$

$= \pi(0 - (-1)) = \pi$

5 $V = \pi \int_1^a y^2\, dx$

$= \pi \int_1^a \dfrac{3}{x}\, dx$

$= \pi [3 \ln x]_1^a$

$= 3\pi \ln a$

$\therefore \pi \ln a^3 = \pi \ln\left(\dfrac{64}{27}\right)$

$\Rightarrow a^3 = \dfrac{64}{27}$

$\Rightarrow a = \dfrac{4}{3}$

6 a Intersections where

$x^2 + 3 = 4x + 3$

$x^2 - 4x = 0$

$x(x - 4) = 0$

$x = 0$ or 4

Substituting into $y = 4x + 3$:

$x = 0 \Rightarrow y = 3$

$x = 4 \Rightarrow y = 4 \times 4 + 3 = 19$

$\therefore$ coordinates of the intersections are $(0, 3)$ and $(4, 19)$

b Volume of revolution of the enclosed region is the difference in the volumes of revolution from the two curves:

$$V = \pi \int_0^4 (4x+3)^2 - (x^2+3)^2 \, dx$$

$$= \pi \int_0^4 16x^2 + 24x + 9 - x^4 - 6x^2 - 9 \, dx$$

$$= \pi \int_0^4 -x^4 + 10x^2 + 24x \, dx$$

$$= \pi \left[-\frac{x^5}{5} + \frac{10x^3}{3} + 12x^2 \right]_0^4$$

$$= \pi \left(-\frac{1024}{5} + \frac{640}{3} + 192 \right)$$

$$= \frac{3008\pi}{15}$$

$$= 630 \ (3SF)$$

> **COMMENT**
> Note that the modulus sign can be dropped because $e^x + 1 > 0$ for all x.

3 $\int_0^m \dfrac{dx}{3x+1} = 1$

$$\left[\frac{1}{3} \ln|3x+1| \right]_0^m = 1$$

$$\frac{1}{3} \ln|3m+1| = 1$$

$$3m + 1 = \pm e^3$$

$$m = \frac{-1 \pm e^3}{3}$$

$$= 6.36 \text{ or } -7.03$$

Reject the negative value, since that would take the integral across $x = -\dfrac{1}{3}$, where the integrand is not defined.

$$\therefore m = \frac{e^3 - 1}{3} = 6.36 \ (3SF)$$

Mixed examination practice 15
Short questions

1 a $\int \dfrac{1}{1-3x} dx = -\dfrac{1}{3} \ln(1-3x) + c$

 b $\int \dfrac{1}{(2x+3)^2} dx = \int (2x+3)^{-2} dx$

$$= -\frac{1}{2}(2x+3)^{-1} + c$$

2 $u = e^x + 1$

$$\Rightarrow \frac{du}{dx} = e^x$$

$$\Rightarrow dx = \frac{du}{e^x}$$

$$\int \frac{e^x}{e^x + 1} dx = \int \frac{e^x}{u} \frac{du}{e^x}$$

$$= \int \frac{1}{u} du$$

$$= \ln|u| + c$$

$$= \ln(e^x + 1) + c$$

4 a Using the product rule:

$$\frac{d}{dx}(x \ln x) = 1 \times \ln x + x \times \frac{1}{x}$$

$$= 1 + \ln x$$

 b $\int \ln x \, dx = \int (1 + \ln x) - 1 \, dx$

$$= x \ln x - x + c$$

5 $V = \pi \int_1^2 \left(\dfrac{1}{x} \right)^2 dx$

$$= \pi \int_1^2 x^{-2} dx$$

$$= \pi \left[-\frac{1}{x} \right]_1^2$$

$$= \frac{\pi}{2}$$

15 Further integration

6 Intersections with the x-axis when

$x \sin x^2 = 0$

$x = 0$ or $\sin x^2 = 0$

$x = 0$ or $x^2 = \pi$

$\therefore x = 0$ or $\sqrt{\pi}$

Area $= \int_0^{\sqrt{\pi}} x \sin x^2 \, dx$

Use substitution: let $u = x^2$; then

$\dfrac{du}{dx} = 2x$

$dx = \dfrac{du}{2x}$

When $x = 0$, $u = 0$

When $x = \sqrt{\pi}$, $u = \pi$

$\therefore \text{Area} = \int_{x=0}^{x=\sqrt{\pi}} x \sin u \, \dfrac{du}{2x}$

$= \dfrac{1}{2} \int_{u=0}^{u=\pi} \sin u \, du$

$= \dfrac{1}{2}[-\cos u]_0^{\pi}$

$= 1$

7 $V = \dfrac{3\pi}{2}$

$\pi \int_0^a (e^x)^2 \, dx = \dfrac{3\pi}{2}$

$\int_0^a e^{2x} \, dx = \dfrac{3}{2}$

$\left[\dfrac{1}{2} e^{2x}\right]_0^a = \dfrac{3}{2}$

$\dfrac{1}{2} e^{2a} - \dfrac{1}{2} = \dfrac{3}{2}$

$e^{2a} = 4$

$a = \dfrac{1}{2} \ln 4 = \ln 2$

8 a $\dfrac{1}{x-2} + \dfrac{5}{(x-2)^2} = \dfrac{x-2}{(x-2)^2} + \dfrac{5}{(x-2)^2}$

$= \dfrac{x-2+5}{(x-2)^2}$

$= \dfrac{x+3}{(x-2)^2}$

b $\int \dfrac{x+3}{(x-2)^2} \, dx = \int \dfrac{1}{x-2} + \dfrac{5}{(x-2)^2} \, dx$

$= \ln(x-2) - \dfrac{5}{x-2} + c$

9 Let $u = \ln x$; then

$\dfrac{du}{dx} = \dfrac{1}{x}$

$dx = x \, du$

$\int \dfrac{1}{x \ln x} \, dx = \int \dfrac{1}{xu} x \, du$

$= \int \dfrac{1}{u} \, du$

$= \ln|u| + c$

$= \ln|\ln x| + c$

Long questions

1 a $I + J = \int \dfrac{\sin x}{\sin x + \cos x} \, dx + \int \dfrac{\cos x}{\sin x + \cos x} \, dx$

$= \int \dfrac{\sin x + \cos x}{\sin x + \cos x} \, dx$

$= \int 1 \, dx$

$= x + c_1$

b $J - I = \int \dfrac{\cos x - \sin x}{\sin x + \cos x} \, dx$

Let $u = \sin x + \cos x$; then

$\dfrac{du}{dx} = \cos x - \sin x$

$dx = \dfrac{du}{\cos x - \sin x}$

$$J - I = \int \frac{\cos x - \sin x}{u} \times \frac{du}{\cos x - \sin x}$$

$$= \int \frac{1}{u} du$$

$$= \ln|u| + c_2$$

$$= \ln|\sin x + \cos x| + c_2$$

c

$$\int \frac{\sin x}{\sin x + \cos x} dx = I$$

$$= \frac{1}{2}((I+J) - (J-I))$$

$$= \frac{1}{2}(x + c_1 - \ln|\sin x + \cos x| - c_2)$$

$$= \frac{1}{2}(x - \ln|\sin x + \cos x|) + c$$

2 a Consider a unit circle about a centre O, with points A and B on the circumference such that $A\hat{O}B = 2\theta \leq \pi$; let M be the midpoint of [AB], so that $A\hat{O}M = M\hat{O}B = \theta$.

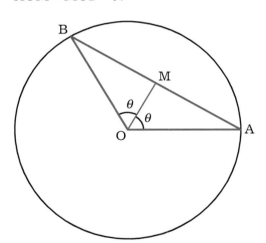

Figure 15ML.2

$AM = AO \sin\theta = \sin\theta$

By the cosine rule,

$AB^2 = AO^2 + BO^2 - 2(AO)(BO)\cos 2\theta$

$\quad = 1 + 1 - 2\cos 2\theta$

$\quad = 2(1 - \cos 2\theta)$

$$AM^2 = \left(\frac{1}{2} AB\right)^2$$

$$= \frac{1}{4} AB^2$$

$$= \frac{1}{2}(1 - \cos 2\theta)$$

i.e. $\sin^2\theta = \frac{1}{2}(1 - \cos 2\theta)$

$1 - \cos^2\theta = \frac{1}{2}(1 - \cos 2\theta)$

$\cos^2\theta = \frac{1}{2}(1 + \cos 2\theta)$

b $\int \cos^2 x \, dx = \frac{1}{2} \int 1 + \cos 2x \, dx$

$$= \frac{1}{2}\left(x + \frac{1}{2}\sin 2x\right) + c$$

$$= \frac{1}{4}(2x + \sin 2x) + c$$

c Substitute $u = 3x$

$$\Rightarrow \frac{du}{dx} = 3$$

$$\Rightarrow dx = \frac{du}{3}$$

$x = 0 \Rightarrow u = 0$

$x = \pi \Rightarrow u = 3\pi$

$$\int_{x=0}^{x=\pi} \cos^2(3x) \, dx = \int_{u=0}^{u=3\pi} \frac{1}{3}\cos^2 u \, du$$

$$= \frac{1}{3} \times \frac{1}{4}[2u + \sin 2u]_0^{3\pi}$$

$$= \frac{\pi}{2}$$

3 a $a = \frac{dv}{dt} = 4t - 10$

$\therefore a(0) = -10 \text{ m s}^{-2}$

b $2t^2 - 10t = 0$

$2t(t-5) = 0$

$t = 0$ or 5

Velocity is zero at times 0 seconds and 5 seconds.

15 Further integration 167

c $h = \int v\, dt$

$= \int 2t^2 - 10t\, dt$

$= \dfrac{2t^3}{3} - 5t^2 + c$

$h(0) = 50$

$\Rightarrow \dfrac{2(0)^3}{3} - 5(0)^2 + c = 50$

$\Rightarrow c = 50$

Closest point to the river occurs when $v = 0$ for the second time (after which she goes up again), i.e. at $t = 5$:

$h(5) = \dfrac{250}{3} - 125 + 50 = \dfrac{25}{3} = 8.33\text{ m}$

d Since $h(5) = \dfrac{25}{3}$ and $h(0) = 50$, distance travelled in the first 5 seconds (moving down) is

$50 - \dfrac{25}{3} = \dfrac{125}{3}\text{ m}$

Since $h(7) = \dfrac{2(7)^3}{3} - 5(7)^2 + 50 = \dfrac{101}{3}$, distance travelled in the next 2 seconds (moving up) is

$\dfrac{101}{3} - \dfrac{25}{3} = \dfrac{76}{3}\text{ m}$

So total distance travelled in the first 7 seconds is

$\dfrac{125}{3} + \dfrac{76}{3} = \dfrac{201}{3} = 67\text{ m}$

e Amy returns to the platform when $h = 50$:

$\dfrac{2t^3}{3} - 5t^2 + 50 = 50$

$2t^3 - 15t^2 = 0$

$t^2(2t - 15) = 0$

$t = 0$ or 7.5

∴ Amy returns to the platform after 7.5 seconds.

16 Summarising data

Exercise 16A

3 The word 'average' is insufficiently clear; half of any population is (equal to or) below the population median, certainly, and if the data is essentially continuous, equality can be considered unfeasible, so that the statement is then true. However, if the data is discrete (e.g. consisting of IQ scores), then it is likely that some children have the average intelligence, and less than half have lower values. If the mean or mode or some other average is being used, there is no reason to think the statement should be true, although it may be.

4 a Sometimes true.

1, 1, 10: median 1, mean 4

1, 10, 10: median 10, mean 7

b Sometimes true.

1, 1, 2: median 1 is in the data set

1, 1, 2, 2: median 1.5 is not in the data set

c Sometimes true.

1, 3: mean 2

1, 2: mean 1.5

d Sometimes true, as in the first example of (c).

e Always true, since the mode must be one of the data values.

f Sometimes true.

1, 1, 2, 3, 4: mode 1, median 2

1, 2, 3, 4, 4: mode 4, median 3

g Always true.

$$\text{Mean} = \frac{x_1 + x_2 + \ldots + x_n}{n} \le \frac{(x_{max} + x_{max} + \ldots + x_{max})}{n}$$
$$= x_{max}$$

h Sometimes true; depends on how many items take the median value.

1, 1, 2, 3, 4: two data items below and two above the median 2

1, 2, 2, 3, 4: one data item below the median 2 and two above it.

5 $\text{Total}_1 = 20.4 \times 14 = 285.6$

$\text{Total}_2 = 16.8 \times 20 = 336$

Grand total $= 285.6 + 336 = 621.6$

Overall mean $= \dfrac{621.6}{34} = 18.3 \, (3\text{SF})$

6 a Require that $\dfrac{3 \times 0.72 + x}{4} \ge 0.75$

$\therefore x \ge 3 - 2.16$

$x \ge 0.84$

She needs to score at least 84% in the fourth paper.

b The highest possible score on the fourth paper is 100%.

$$\frac{3 \times 0.72 + 1}{4} = 0.79$$

So she can score at most 79% overall.

7 $x + y + 1 + 3 + 10 = 5 \times 5.4 = 27$

$\Rightarrow x + y = 13$

The median is 5, and the median of five values must be one of the values.

$\therefore x$ and y are 5 and 8.

8 The mode is 2, which has a frequency of 10, so $p < 10$.

Total number of students is $23 + p$. For the median to be 3 marks, $15 < \dfrac{23 + p}{2}$

$\therefore 30 < 23 + p$

$\Rightarrow 7 < p$

Hence $7 < p < 10$

So p can be 8 or 9.

9 If 'average' can take different meanings, there are numerous ways this can be true.

If 'average' represents 'mean' throughout, then it can still be true, provided the number of attempts varies.

For example:

Amy plays level 1 twice and scores 90 and 100 (mean 95).

Bob plays level 1 four times and scores 90, 91, 93 and 94 (mean 92).

Amy plays level 2 twice and scores 50 and 60 (mean 55).

Bob plays level 2 twice and scores 42 and 52 (mean 47).

Clearly Amy's average is higher than Bob's in both level 1 and level 2.

However, their overall averages are

Amy: $\dfrac{90 + 100 + 50 + 60}{4} = 75$

Bob: $\dfrac{90 + 91 + 93 + 94 + 42 + 52}{6} = 77$

So Bob has a higher overall average.

Exercise 16B

2 a False: the mean must be between 20 and 32, but the range is 12.

b Impossible to be sure; could be true, for example 20, 26, 26, 26, 26, 32.

c True: the mean must be between 20 and 32, so greater than 10.

d Impossible to be sure; could be true, for example 20, 21, 21, 22, 28, 32

mean $= \dfrac{20 + 21 + 21 + 22 + 28 + 32}{6} = 24,$

mode $= 21$

e Impossible to be sure; could be true, for example 20, 21, 22, 23, 24, 43

median $= \dfrac{22 + 23}{2} = 22.5$

3 There are 8 data values.

The median is $\dfrac{5 + 6}{2} = 5.5$

Lower data set 3, 4, 5, 5 has median

$\dfrac{4 + 5}{2} = 4.5$, so $Q_1 = 4.5$

Upper data set 6, 8, 11, 13 has median

$\dfrac{8 + 11}{2} = 9.5$, so $Q_3 = 9.5$

$IQR = Q_3 - Q_1 = 5$

4 a There are 7 data values.

The median is 8

Lower data set 5, 5, 7 has median 5, so $Q_1 = 5$

Upper data set 9, x, 13 has median x, so $Q_3 = x$

$IQR = Q_3 - Q_1 = 7$

$\therefore x - 5 = 7$

$\Rightarrow x = 12$

b From GDC, for the data set 5, 5, 7, 9, 12, 13: $\sigma = 2.92$

5 Median is $b = 12$

Range is $c - a = 12$

$\Rightarrow c = a + 12$

Mean is $\dfrac{a+b+c}{3} = 14$

$\Rightarrow a+b+c = 42$

$a + 12 + c = 42$

$\Rightarrow a + c = 30$

$\therefore 2a + 12 = 30$

$\Rightarrow a = 9,\ b = 12,\ c = 21$

Exercise 16C

2 a The youngest could be 17 years exactly (i.e. 17th birthday on the day the data is collected); the oldest could be one day short of 21 years old (i.e. 21st birthday the day after the data is collected).

b The least is 17 pencils; the most is 20 pencils.

c The lower bound is 16.5 cm; the upper bound is 20.5 cm.

d The lower bound is $16.01 per hour; the upper bound is $20.00 per hour

4 a There are 50 boxes, so the median will be the mean of the 25th and 26th boxes:

$\text{median} = \dfrac{1+2}{2} = 1.5$ broken eggs

b $\text{Mean} = \dfrac{(17\times 0)+(1\times 8)+(2\times 7)+(3\times 7)+(4\times 6)+(5\times 5)+(6\times 0)}{50}$

$= 1.84$ broken eggs

5 Range $= q - 5 = 15 \Rightarrow q = 20$

Mean $= \dfrac{(5 \times 6) + 10p + 15 \times 2p + q \times 2}{8 + 3p} = 12.6$

$\therefore 30 + 40p + 40 = 12.6 \times (8 + 3p)$

$2.2p = 30.8$

$\Rightarrow p = 14$

6 Range $= 20 \Rightarrow q = 0$ (otherwise range would be 40)

Mean $= \dfrac{12 \times 20 + 40p + 60q}{12 + p + q} = 30.4$

$\therefore 240 + 40p = 30.4(12 + p)$

$240 + 40p = 364.8 + 30.4p$

$9.6p = 124.8$

$\Rightarrow p = 13$

Exercise 16D

3 a Median will be at cumulative frequency 40, which corresponds to 197 seconds.

b Q_1 will be at cumulative frequency 20, which corresponds to 191 seconds

Q_3 will be at cumulative frequency 60, which corresponds to 203 seconds

$\therefore IQR = Q_3 - Q_1 = 12$ seconds

c The middle 50% is that cohort lying between Q_1 and Q_3, so $c = 191$ and $d = 203$

4 a 30 minutes corresponds to a cumulative frequency of 56

50 minutes corresponds to a cumulative frequency of 162

So approximately 106 students take between 30 and 50 minutes to travel to school, which is 53% of the population.

b 20% will be at cumulative frequency 40, which corresponds to 28 minutes

$\therefore$ 80% of students take more than 28 minutes to travel to school.

5 Median $= 12$

$IQR = 18 - 9 = 9$

6 a Estimate of the mean uses the midpoints of the categories: 149.5, 249.5, …

$$\text{Mean} = \frac{1}{200}[(149.5 \times 12)+(249.5 \times 36)+(349.5 \times 42)+(449.5 \times 53)$$
$$+(549.5 \times 33)+(649.5 \times 20)+(749.5 \times 4)]$$
$$= 417 \text{ pages}$$

b TABLE 16D.6

No. pages	199	299	399	499	599	699	799
Cumulative frequency	12	48	90	143	176	196	200

c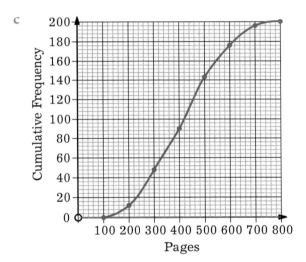

Figure 16D.6

d Median will be at cumulative frequency 100, which corresponds to 420 pages.

e Q_1 will be at cumulative frequency 50, which corresponds to 310 pages

Q_3 will be at cumulative frequency 150, which corresponds to 520 pages

$IQR = Q_3 - Q_1 = 210$ pages

f 450 pages corresponds to cumulative frequency 116.
Approximately 84 books have more than 450 pages, representing about 42% of the population.

Exercise 16E

2 The gradient of the cumulative frequency curve corresponds to the frequency value in that interval.

Curve A is steep in the middle and less steep at either end, so corresponds to histogram 3.

Curve B is steepest at low data values and gradually becomes less steep, so corresponds to histogram 2.

Curve C is shallow then steep then less steep, then steep again and finally becomes shallow, so corresponds to histogram 1, which is bimodal.

3 The data table is as follows:

TABLE 16E.3

Value	$0 \leq x < 5$	$5 \leq x < 10$	$10 \leq x < 15$	$15 \leq x < 20$	$20 \leq x < 25$
Midpoint	2.5	7.5	12.5	17.5	22.5
Frequency	20	15	7	16	4

Estimate of mean $= \dfrac{(2.5 \times 20)+(7.5 \times 15)+(12.5 \times 7)+(17.5 \times 16)+(22.5 \times 4)}{20+15+7+16+4} = 10$

Estimate of σ is 6.71 (from GDC)

Exercise 16F

3 a Mean $= \dfrac{(x-2)+x+(x+1)+(x+5)}{4}$

$= \dfrac{4x+4}{4}$

$= x+1$

b New mean is $x+1-6 = x-5$

4 1 mpg = 0.354 kpl

Median $= 32 \times 0.354 = 11.3$ kpl

Variance $= 60 \times 0.354^2 = 7.52$ kpl²

5 $F = 1.8C + 32$

a $\bar{F} = 1.8\bar{C} + 32$

$= 1.8 \times 4 + 32$

$= 39.2$

b $IQR_C = \dfrac{IQR_F}{1.8}$

$= \dfrac{2}{1.8}$

$= 1.11$

6 $y = ax - b$

$\therefore \text{median}_y = a \times \text{median}_x - b$

$= 20a - b$

$IQR_y = a \times IQR_x = 10a$

Require that $20a - b = 10a$

$\therefore b = 10a$

7 Multiplying all data values by a negative number $-k$ would spread the data out as much as multiplying by the positive number k, so the general rule should be:

If you multiply (or divide) every data item by the value x, all direct measures of the spread of the data will be multiplied (or divided) by $|x|$.

Exercise 16G

3 a From GDC, $r = 0.990$

b There is a (very) strong positive correlation between processor speed and life expectancy.

c Not at all, nor even vice versa. Correlation does not imply causation; a more reasonable explanation would be that both measures have increased over the 20 years of the data due to gains in scientific knowledge, both medical and technological.

4 a From GDC, $r = 0.876$

b There is a strong positive correlation between a car's age and its braking distance.

c The statement is true; correlation does indicate a tendency for one variable to vary in line with the other.

5 a From GDC, $r = 0.162$

b The correlation is very weak, so this data would be poor evidence to support Gavin's claim. Moreover, Gavin is asserting that years in education drives income, but correlation cannot be considered to be causation; there could be other factors driving both variables.

6 a From GDC, $r = -0.996$

b From GDC, $r = -0.996$

c The two correlation coefficients are the same, because Fahrenheit and Celsius temperatures are linearly related ($F = 1.8C + 32$), and so the correlation of either with reaction time must have the same value.

Exercise 16H

2 a The regression line passes through $(\bar{d}, \bar{n})$

$\therefore \bar{n} = 0.14\bar{d} - 20$

$\bar{d} = \dfrac{\bar{n} + 20}{0.14} = \dfrac{36 + 20}{0.14} = 400$

The mean number of calories in a dessert is 400.

b The median number of calories cannot be determined from the regression line; we may expect it to be close to $\dfrac{32 + 20}{0.14} = 371$, but without knowing the actual calorie values of each dessert, we cannot make an accurate calculation of the median.

3 **a** From GDC, $r = 0.962$

 b The independent variable is d, whose values are fixed within the experiment.

 v is the dependent variable, since this is the value being measured for the fixed values of d.

 c From GDC, the regression line is
 $$v = 0.275d + 11.5$$
 $$\therefore v(45) = 0.275 \times 45 + 11.5$$
 $$= 23.9 \text{ km/h}$$

 d The data set contains only measurements for d between 10 and 60. Estimating beyond the data set (extrapolating) is not reliable.

4 **a** From GDC, $r = -0.700$

 b Temperature T is the independent variable, whose values are fixed within the experiment.

 c From GDC, the regression line is
 $$h = 24.6 - 0.0236T$$

 d The correlation is not very strong, meaning that any prediction would be unreliable.

 Additionally, $h = 20$ cm is outside the range of data values, so making such a prediction would be extrapolation; in fact, since the gradient of the regression line is negative, extrapolation would be asserting that the lower the temperature, the higher the cake will rise, which is plainly nonsense for low temperatures!

 A scatter plot of the data shows that height peaks around 340 degrees at 18.1 cm, so it is perhaps not even possible to reach 20 cm!

5 **a** False; it means only that there is no linear relationship. The diagram on page 493 of the coursebook shows the plot of data exactly following a quadratic curve, but $r = 0$ for that data set, although clearly the two variables are related, and predictions may accordingly be made.

 b False. If there are at least two data points, then $k > 0 \Rightarrow r = 1$; however, $k < 0 \Rightarrow r = -1$, and if $k = 0$ then $r = 0$.

 c True.

 d False. As r increases, the strength of the correlation rises (for positive r) or falls (for negative r), but the gradient of the regression line is not predicted by the correlation coefficient.

6 **a** **i** From GDC, $r = -0.328$

  ii From GDC, $r = 0.996$

 b From GDC, $y = 1.11x^2 - 3.55$

Mixed examination practice 16

Short questions

1 **a** Ordered data: 108, 111, 115, 122, 124, 127, 135, 139, 140

 There are 9 data items.

 The median is 124

 Lower data set 108, 111, 115, 122 has median 113, so $Q_1 = 113$

 Upper data set 127, 135, 139, 140 has median 137, so $Q_3 = 137$

 $IQR = Q_3 - Q_1 = 24$

b Mean = $\dfrac{108+111+115+122+124+127+135+139+140}{9}$

   $= 125$

c From GDC, variance = 124 (3SF)

2 a α: Most data is at the right end of the range – histogram D

β: Data is concentrated in the centre and symmetrical – histogram B

γ: Data is distributed away from the centre and symmetrical – histogram C

b The gradient of the cumulative frequency curve corresponds to the frequency value in that interval.

Curve (i) is steep in the middle, i.e. the data is concentrated in the centre, and is also symmetrical – histogram B

Curve (ii) has constant gradient, so the data is evenly distributed – histogram A

Curve (iii) is steep at either end and has smaller gradient at the centre, i.e. the data is distributed away from the centre, and it is symmetrical – histogram C

4 a 40 marks corresponds to cumulative frequency 100, so approximately 100 students scored 40 marks or less on the test.

b Q_1 will be at cumulative frequency 200, which corresponds to 55 marks

Q_3 will be at cumulative frequency 600, which corresponds to 75 marks

So $a = 55$ and $b = 75$

4 a $p = 10f - 100$

b $\bar{p} = 10\bar{f} - 100$

   $= 10 \times 83 - 100$

   $= 730$ pounds

c $\text{Var}(p) = 10^2 \times \text{Var}(f)$

   $= 6000$ pounds^2

5 Median $= b = 15$

Range $= c - a = 10$

$\Rightarrow c = a + 10$

Mean $= \dfrac{a+b+c}{3} = 13$

16 Summarising data

∴ $a+15+c=39$

⇒ $a+c=24$

∴ $2a+10=24$

⇒ $a=7, b=15, c=17$

Long questions

1 a The median will be at cumulative frequency 15, which corresponds to 11 cm.

b From the cumulative frequency diagram:

TABLE 16ML.1

Height (h)	Frequency
$0 < h \leq 5$	4
$5 < h \leq 10$	$13 - 4 = 9$
$10 < h \leq 15$	$21 - 13 = 8$
$15 < h \leq 20$	$26 - 21 = 5$
$20 < h \leq 25$	$30 - 26 = 4$

c Estimate of the mean uses midpoint values of the intervals:

Estimate of mean

$$= \frac{(2.5 \times 4)+(7.5 \times 9)+(12.5 \times 8)+(17.5 \times 5)+(22.5 \times 4)}{30}$$

$$= \frac{355}{30} = 11.8 \text{ cm (3SF)}$$

2 a **TABLE 16ML.2**

Mark	≤ 10	≤ 20	≤ 30	≤ 40	≤ 50	≤ 60	≤ 70	≤ 80	≤ 90	≤ 100
Number	15	65	165	335	595	815	905	950	980	1000

b

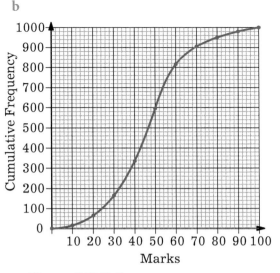

Figure 16ML.2

c i Median will be at cumulative frequency 500, which corresponds to 46 marks.

   ii 34 marks corresponds to cumulative frequency 220, so 220 students scored less than 35 marks.

   iii The top 15% will be at cumulative frequency 850 and above, which corresponds to 63 marks and above.

   So 63 marks or above are needed to be awarded a distinction.

3 a From GDC, $r = -0.660$

   b The day of week might have a significant effect, as would the time of year (which while related to temperature is nonetheless a separate factor).

   c Temperature might influence sales, but sales could not influence the temperature. Temperature is therefore the independent variable.

   d From GDC, the regression line is
      $S = 617 - 19.0\,T$

   e From the regression line, the sales at 0°C would be approximately £617.

   f The correlation is not particularly strong; and in any case, −20°C is well outside the range of temperature data, and extrapolation is unreliable since there is nothing to suggest that a linear relationship would extend that far.

4 a From GDC, $r = 0.945$

   b From GDC, the regression line is
      $n = 5.00T - 87.8$

   An estimate for n when $T = 25$ would be $5.00 \times 25 - 87.8 = 37.2$ ice creams.

   c Zero ice cream sales does not lie within the observed data, so any estimate for this would be extrapolation, which is unreliable and would likely fall outside the context of the data collected.

   Also, 'sales' is the dependent variable in this data set, and although the correlation coefficient is high, it is not strictly appropriate to use the dependent variable to predict the independent variable on a regression line.

17 Probability

Exercise 17A

3 Total number of cases:
$20 + 68 + 10 + 30 = 128$

Total number of positive results:
$18 + 68 = 86$

$\therefore P(\text{positive result}) = \dfrac{86}{128} = 0.672$

4 That depends on the relative numbers of light bulbs in each trial.

If each trial tested the same number of bulbs, then it would be true that, overall, 50% would pass.

Otherwise, this is not the case. For example: $n_1 = 50$, of which 20 pass; $n_2 = 100$, of which 60 pass.
Then $n = 50$ in total, of which 80 pass, giving an overall pass rate of 53%.

Exercise 17B

4 For a fair six-sided die, $P(\text{Odd}) = 0.5$ and $P(\text{Prime}) = 0.5$
So $P(\text{Odd}) + P(\text{Prime}) = 1$
But Odd is not the complement of Prime (3 and 5 are in both, 4 and 6 are in neither).

7 Draw out the table of the event space:

TABLE 17B.7

		Die A				
Die B	1	2	3	4	5	6
1	1	1	1	1	1	1
2	1	2	1	2	1	2
3	1	1	3	1	1	3
4	1	2	1	4	1	2
5	1	1	1	1	5	1
6	1	2	3	2	1	6

$P(\text{HCF} = 1) = \dfrac{23}{36}$

So expected number of '1' scores in 180 throws is

$180 \times \dfrac{23}{36} = 115$

8 Cases for score less than 6 (i.e. 5 or less):

Total 3: $(1,1,1)$

Total 4: $(1,1,2), (1,2,1), (2,1,1)$

Total 5: $(1,1,3), (1,2,2), (1,3,1), (2,1,2), (2,2,1), (3,1,1)$

This gives 10 cases out of the total $6^3 = 216$ possible outcomes.

$\therefore P(\text{score} < 6) = \dfrac{10}{216} = \dfrac{5}{108}$

Exercise 17C

4

> **COMMENT**
> A Venn diagram is sufficient working; the calculations needed to populate the diagram are given first but do not need to be explicitly shown in an examination answer. Below the diagram a stand-alone algebraic method is given, which could be used instead of a diagram.

$P(S \cap L') = P(S) - P(S \cap L)$
$= 60\% - 50\%$
$= 10\%$

$P(L \cap S') = P(L) - P(L \cap S)$
$= 85\% - 50\%$
$= 35\%$

$P(L' \cap S') = 1 - P(L) - P(S \cap L')$
$= 100\% - 60\% - 35\%$
$= 5\%$

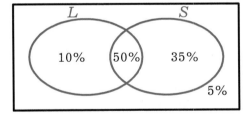

Figure 17C.4

$P(L' \cap S') = P\big((L \cup S)'\big)$
$= 1 - (P(L) + P(S) - P(S \cap L))$
$= 1 - P(L) - P(S) + P(S \cap L)$
$= 100\% - 85\% - 60\% + 50\%$
$= 5\%$

5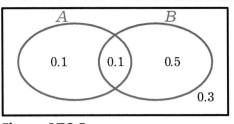

Figure 17C.5

$P(B \cap A') = P(A \cup B) - P(A)$
$= 0.7 - 0.2 = 0.5$

$P(B') = 1 - P(B)$
$= 1 - (P(A \cap B) + P(A' \cap B))$
$= 1 - (0.1 + 0.5)$
$= 0.4$

6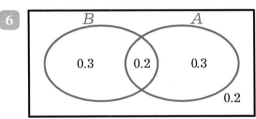

Figure 17C.6

$P\big((A \cup B)'\big) = P(A' \cap B') = 0.2$

$P(A \cap B') = P(B') - P(A' \cap B')$
$= 1 - P(B) - P(A' \cap B')$
$= 1 - 0.5 - 0.2$
$= 0.3$

7 a $\dfrac{1000}{6} = 166.7$

so there are 166 multiples of 6 among the first 1000 numbers.

$\therefore P(\text{multiple of } 6) = \dfrac{166}{1000} = 0.166$

b The numbers which are multiples of both 6 and 8 are multiples of $\text{LCM}(6, 8) = 24$

$\dfrac{1000}{24} = 41.7$

so there are 41 multiples of 24 among the first 1000 numbers

$\therefore P(\text{multiple of } 24) = \dfrac{41}{1000} = 0.041$

Exercise 17D

> **COMMENT**
>
> In all these questions a tree diagram, populated with relevant values, is sufficient preliminary working. The tree diagrams are given in these answers, together with full algebraic working such as would be needed for an answer if a tree diagram were not drawn.

2

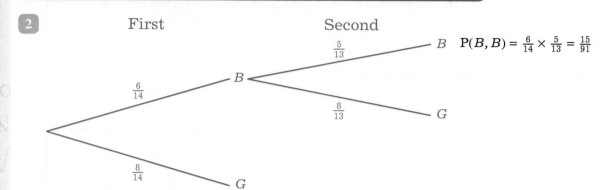

Figure 17D.2

B = boy; G = girl

$$P(B,B) = \frac{6}{14} \times \frac{5}{13} = \frac{15}{91}$$

3

> **COMMENT**
>
> In a question like this, there is no need to draw a 'full' tree, since only a few results are of interest. Thus, after blue in the first draw, all that is of interest is whether the second ball is green or not, so G/G' branches are sufficient – there is no need for three branches $B/G/R$. Similarly, after G in the first draw only B/B' branches are needed, and a first draw of R need not be detailed further.

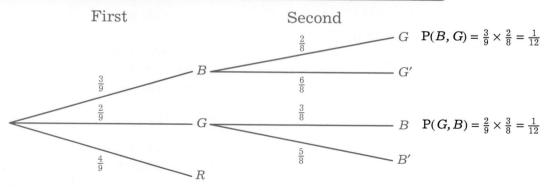

Figure 17D.3

B = blue; G = green; R = red

$$P(\text{Blue and Green}) = P(B,G) + P(G,B)$$
$$= \frac{3}{9} \times \frac{2}{8} + \frac{2}{9} \times \frac{3}{8}$$
$$= \frac{1}{12} + \frac{1}{12}$$
$$= \frac{1}{6}$$

4

```
                    2/9
              X <
        1/3     \   7/9    Y'  P(X ∩ Y') = 1/3 × 7/9 = 7/27
      /              \
     /                 Y   P(X' ∩ Y) = 2/3 × 1/3 = 6/27
      \         1/3  /
        2/3    /
              X' <
                    2/3
                         Y'  P(X' ∩ Y') = 2/3 × 2/3 = 12/27
```

Figure 17D.4

a $P(Y') = P(X \cap Y') + P(X' \cap Y')$
$\quad = P(Y'|X)P(X) + P(Y'|X')P(X')$
$\quad = \frac{7}{9} \times \frac{1}{3} + \frac{2}{3} \times \frac{2}{3}$
$\quad = \frac{19}{27}$

b $P(X' \cap Y') = P(Y') + P(Y \cap X')$
$\quad\quad\quad\quad = P(Y') + P(Y|X')P(X')$
$\quad\quad\quad\quad = \frac{19}{27} + \frac{1}{3} \times \frac{2}{3}$
$\quad\quad\quad\quad = \frac{25}{27}$

Alternatively:
$P(X' \cup Y') = 1 - P(X \cap Y)$
$\quad\quad\quad\quad = 1 - P(Y|X)P(X)$
$\quad\quad\quad\quad = 1 - \frac{2}{9} \times \frac{1}{3} = \frac{25}{27}$

> **COMMENT**
> Normally this alternative of calculating the complement to the union would be the faster calculation, but in this question we can with equal ease harness the answer to part (a).

5

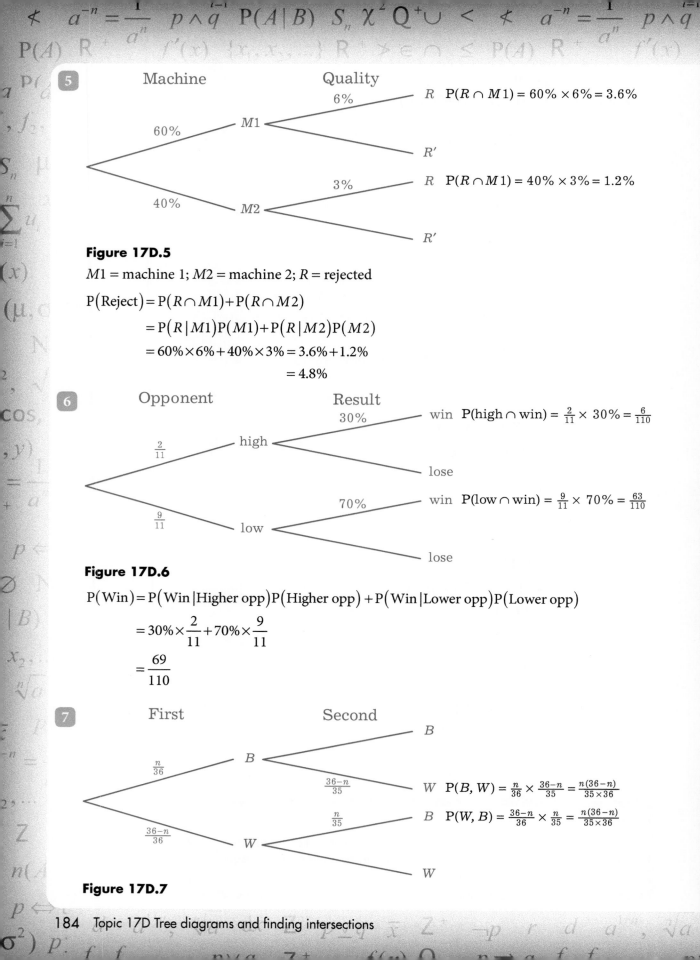

Figure 17D.5

M1 = machine 1; M2 = machine 2; R = rejected

$$P(\text{Reject}) = P(R \cap M1) + P(R \cap M2)$$
$$= P(R|M1)P(M1) + P(R|M2)P(M2)$$
$$= 60\% \times 6\% + 40\% \times 3\% = 3.6\% + 1.2\%$$
$$= 4.8\%$$

6

Figure 17D.6

$$P(\text{Win}) = P(\text{Win}|\text{Higher opp})P(\text{Higher opp}) + P(\text{Win}|\text{Lower opp})P(\text{Lower opp})$$
$$= 30\% \times \frac{2}{11} + 70\% \times \frac{9}{11}$$
$$= \frac{69}{110}$$

7

Figure 17D.7

184 Topic 17D Tree diagrams and finding intersections

B = black; W = white

Let the number of black counters be n.

If P(Same) = P(Different) then P(Same) = $\frac{1}{2}$

P(Same) = P(B,B) + P(W,W)

$$\therefore \frac{n}{36} \times \frac{n-1}{35} + \frac{36-n}{36} \times \frac{35-n}{35} = \frac{1}{2}$$

$n(n-1) + (36-n)(35-n) = 630$

$2n^2 - 72n + 630 = 0$

$n^2 - 36n + 315 = 0$

$(n-15)(n-21) = 0$

$n = 15$ or 21

Exercise 17E

3 a For independent events A and B, $P(A \cap B) = P(A) \times P(B)$

$P(A \cup B) = P(A) + P(B) - P(A \cap B)$

$\therefore 0.72 = 0.6 + P(B) - 0.6P(B)$

$0.4P(B) = 0.12$

$\Rightarrow P(B) = 0.3$

b $P(A \cap B') + P(A' \cap B) = P(A \cup B) - P(A \cap B)$

$= 0.72 - 0.6 \times 0.3$

$= 0.54$

4 a $P(T \cap S) = P(T) \times P(S)$

$= 92\% \times 68\%$

$= 62.56\%$

b $P(T' \cap S') = P(T') \times P(S')$

$= 8\% \times 32\%$

$= 2.56\%$

c The event 'at least one working' is the complement of 'neither working':

$P(T \cup S) = 1 - P(T' \cap S')$

$= 97.44\%$

5 a $P(\text{all different}) = 1 \times \frac{7}{8} \times \frac{6}{8} \times \frac{5}{8}$

$= \frac{105}{256} = 0.410$

> **COMMENT**
>
> This calculation assumes that there is an effectively unlimited number of each type of toy, which is implicit in the way the question is phrased. If there were limited numbers, the probabilities would change to reflect this – after all, in the most extreme case, if there were only 8 packets of crisps in the world and each had a different toy, David would have complete certainty that he would get a different toy each time.

b Using the complement: let p be the probability that he fails to get any gyroscopes or yo-yos; then

$$p = \left(\frac{6}{8}\right)^4 = \frac{81}{256} = 0.316$$

$$\therefore \text{P(at least one gyroscope or yo-yo)} = 1 - \frac{81}{256}$$

$$= \frac{175}{256}$$

$$= 0.684$$

6 For independent events A and B, $\text{P}(B) = \dfrac{\text{P}(A \cap B)}{\text{P}(A)}$

$$\text{P}(A \cup B) = \text{P}(A) + \text{P}(B) - \text{P}(A \cap B)$$

$$\text{P}(A) = \text{P}(A \cap B) + \text{P}(A \cap B')$$

$$= 0.3 + 0.6 = 0.9$$

$$\therefore \text{P}(B) = \frac{0.3}{0.9} = \frac{1}{3}$$

Hence $\text{P}(A \cup B) = 0.9 + \dfrac{1}{3} - 0.3$

$$= \frac{28}{30} = \frac{14}{15}$$

$$= 0.933$$

Exercise 17F

3 a i $\text{P}(A \cap B) = \text{P}(A) + \text{P}(B) - \text{P}(A \cup B)$

$$= 0.6 + 0.2 - 0.7$$

$$= 0.1$$

ii $P(A) \times P(B) = 0.12 \neq P(A \cap B)$
∴ A and B are not independent.

b $P(B|A) = \dfrac{P(A \cap B)}{P(A)}$

$= \dfrac{0.1}{0.6}$

$= \dfrac{1}{6}$

4 a $P(A \cap B) = P(B|A) \times P(A)$

$= \dfrac{2}{3} \times \dfrac{1}{2}$

$= \dfrac{1}{3}$

b $P(B) = P(A \cup B) - P(A) + P(A \cap B)$

$= \dfrac{4}{5} - \dfrac{2}{3} + \dfrac{1}{3}$

$= \dfrac{7}{15}$

c $P(A) \times P(B) = \dfrac{2}{3} \times \dfrac{7}{15} = \dfrac{14}{45} \neq P(A \cap B)$

∴ A and B are not independent

5

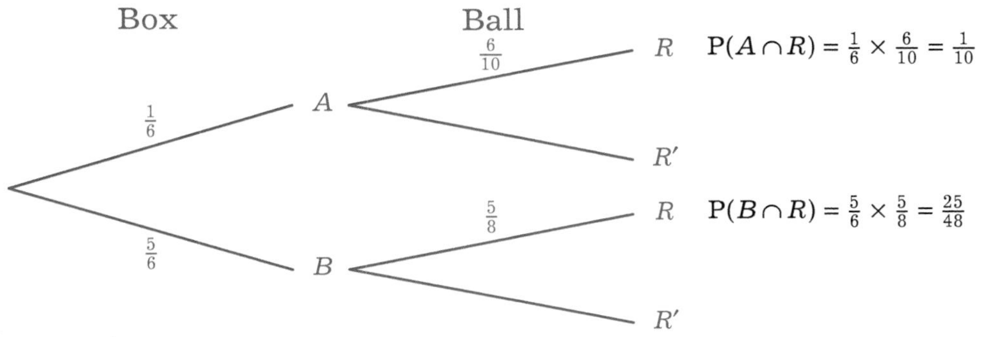

Box: A (1/6), B (5/6)
Ball from A: R (6/10), R'
Ball from B: R (5/8), R'

$P(A \cap R) = \dfrac{1}{6} \times \dfrac{6}{10} = \dfrac{1}{10}$

$P(B \cap R) = \dfrac{5}{6} \times \dfrac{5}{8} = \dfrac{25}{48}$

Figure 17F.5

A = box A; B = box B; R = red

a $P(R) = P(R|A) \times P(A) + P(R|B) \times P(B)$

$= \dfrac{6}{10} \times \dfrac{1}{6} + \dfrac{5}{8} \times \dfrac{5}{6}$

$= \dfrac{1}{10} + \dfrac{25}{48}$

$= \dfrac{149}{240}$

$= 0.621$

b $P(B|R) = \dfrac{P(B \cap R)}{P(R)}$

$= \dfrac{P(R|B)P(B)}{P(R)}$

$= \dfrac{\frac{5}{8} \times \frac{5}{6}}{\frac{149}{240}}$

$= \dfrac{125}{149}$

$= 0.839$

6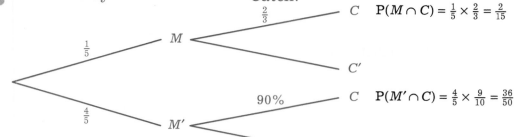

Figure 17F.6

C = Robert catches the train; M = Monday

a $P(C) = P(C|M) \times P(M) + P(C|M') \times P(M')$

$= \dfrac{2}{3} \times \dfrac{1}{5} + \dfrac{9}{10} \times \dfrac{4}{5}$

$= \dfrac{2}{15} + \dfrac{18}{25}$

$= \dfrac{64}{75}$

$= 0.853$

b $P(M|C) = \dfrac{P(M \cap C)}{P(C)}$

$= \dfrac{\frac{2}{15}}{\frac{64}{75}}$

$= \dfrac{5}{32}$

$= 0.156$

7

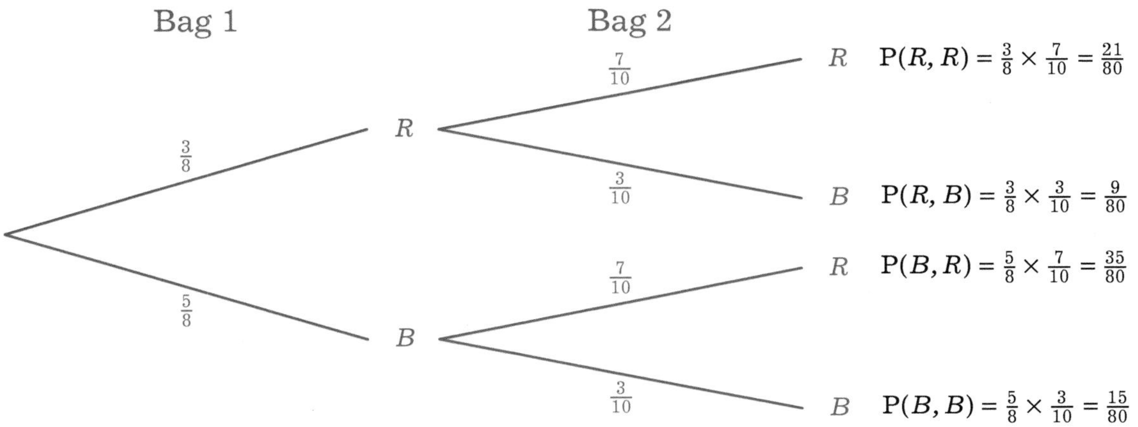

Bag 1   Bag 2

Tree diagram:
- Bag 1: R with probability $\frac{3}{8}$; B with probability $\frac{5}{8}$
- From R (Bag 2): R with $\frac{7}{10}$ → $P(R,R) = \frac{3}{8} \times \frac{7}{10} = \frac{21}{80}$
- From R (Bag 2): B with $\frac{3}{10}$ → $P(R,B) = \frac{3}{8} \times \frac{3}{10} = \frac{9}{80}$
- From B (Bag 2): R with $\frac{7}{10}$ → $P(B,R) = \frac{5}{8} \times \frac{7}{10} = \frac{35}{80}$
- From B (Bag 2): B with $\frac{3}{10}$ → $P(B,B) = \frac{5}{8} \times \frac{3}{10} = \frac{15}{80}$

Figure 17F.7

R = red; B = blue

a $P(\text{same colour}) = P(R,R) + P(B,B)$

$$= \frac{6}{16} \times \frac{7}{10} + \frac{10}{16} \times \frac{3}{10}$$

$$= \frac{21}{80} + \frac{15}{80}$$

$$= \frac{9}{20}$$

$$= 0.45$$

b $P(R1 \mid \text{different}) = \dfrac{P(R1 \cap \text{different})}{P(\text{different})}$

$$= \frac{P(R1 \cap B2)}{P(R1 \cap B2) + P(B1 \cap R2)}$$

$$= \frac{\frac{6}{16} \times \frac{3}{10}}{\frac{6}{16} \times \frac{3}{10} + \frac{10}{16} \times \frac{7}{10}}$$

$$= \frac{\frac{9}{80}}{\frac{9}{80} + \frac{35}{80}}$$

$$= \frac{9}{44}$$

$$= 0.205$$

8

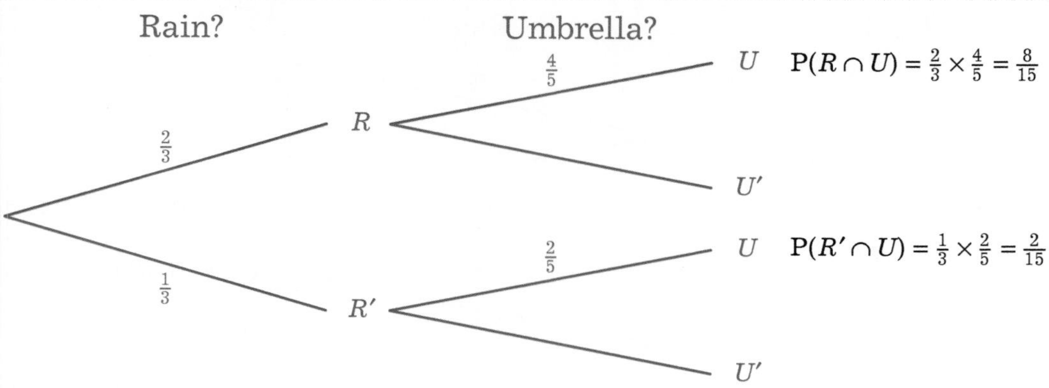

Rain? Umbrella?

[Tree diagram: R (2/3) → U (4/5), U'; R' (1/3) → U (2/5), U']

$P(R\cap U) = \frac{2}{3}\times\frac{4}{5} = \frac{8}{15}$

$P(R'\cap U) = \frac{1}{3}\times\frac{2}{5} = \frac{2}{15}$

Figure 17F.8

R = raining; U = bring umbrella

a $\quad P(U) = P(U\cap R) + P(U\cap R')$
$= P(U|R)P(R) + P(U|R')P(R')$
$= \dfrac{4}{5}\times\dfrac{2}{3} + \dfrac{2}{5}\times\dfrac{1}{3}$
$= \dfrac{10}{15}$
$= \dfrac{2}{3}$

b $\quad P(R|U) = \dfrac{P(R\cap U)}{P(U)}$
$= \dfrac{P(U|R)P(R)}{P(U)}$
$= \dfrac{\frac{8}{15}}{\frac{2}{3}}$
$= \dfrac{4}{5}$

9

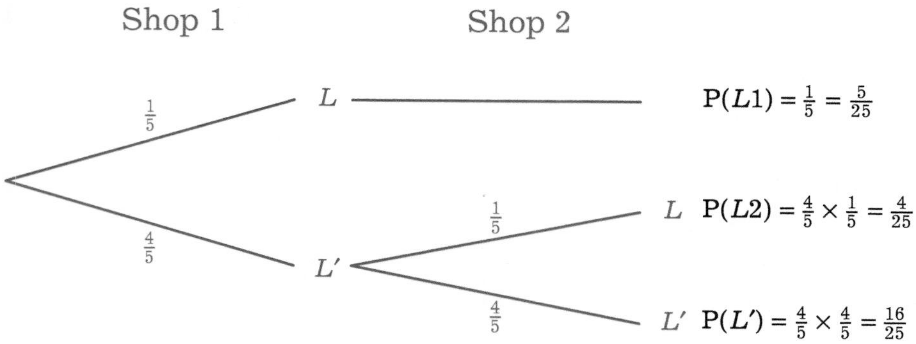

Shop 1 Shop 2

[Tree diagram: L (1/5) → L; L' (4/5) → L (1/5), L' (4/5)]

$P(L1) = \frac{1}{5} = \frac{5}{25}$

$P(L2) = \frac{4}{5}\times\frac{1}{5} = \frac{4}{25}$

$P(L') = \frac{4}{5}\times\frac{4}{5} = \frac{16}{25}$

Figure 17F.9

L = leaves umbrella in shop

$$P(L2|L) = \frac{P(L2)}{P(L1)+P(L2)}$$

$$= \frac{\frac{4}{5} \times \frac{1}{5}}{\frac{1}{5} + \frac{4}{5} \times \frac{1}{5}}$$

$$= \frac{4}{9}$$

10 $P(A \cup B) = P(A) + P(B) - P(A \cap B)$
$\qquad = P(A) + P(B) - P(A|B)P(B)$

$\therefore \dfrac{4}{5} = \dfrac{2}{3} + P(B) - \dfrac{1}{5}P(B)$

$\dfrac{4}{5}P(B) = \dfrac{4}{5} - \dfrac{2}{3} = \dfrac{2}{15}$

$\Rightarrow P(B) = \dfrac{1}{6}$

Exercise 17G

> **COMMENT**
>
> For the questions in this section, the focus should be on completing the diagram, after which subsequent values can be read off with little or no working shown. Preliminary calculations for completing the diagram are given in each question to explain how the diagram is filled in, but usually you would not need to show these calculations in an examination.

1 a F = football; B = badminton

$n(\text{F only}) = n(F) - n(F \cap B) = 34 - 5 = 29$

$n(\text{B only}) = n(B) - n(F \cap B) = 18 - 5 = 13$

$n(F' \cap B') = 145 - 29 - 13 - 5 = 98$

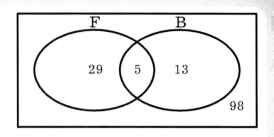

Figure 17G.1

b From diagram: $n(F' \cap B') = 98$

c $P(B) = \dfrac{18}{145}$

d $P(B|F) = \dfrac{P(B \cap F)}{P(F)}$

$\qquad = \dfrac{n(B \cap F)}{n(F)}$

$\qquad = \dfrac{5}{34}$

2 a M = mathematics; E = economics

$n(M \cup E) = 145 - 72 = 73$

$n(M \cap E) = n(M) + n(E) - n(M \cup E)$
$\qquad = 58 + 47 - 73$
$\qquad = 32$

$n(\text{M only}) = n(M) - n(M \cap E)$
$\qquad = 58 - 32$
$\qquad = 26$

$n(\text{E only}) = n(E) - n(M \cap E)$
$\qquad = 47 - 32$
$\qquad = 15$

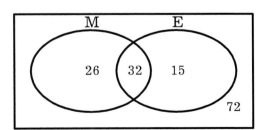

Figure 17G.2

b From diagram: $n(M \cap E) = 32$

c $P(M \cap E | M) = \dfrac{P(M \cap E)}{P(M)}$

$= \dfrac{n(M \cap E)}{n(M)}$

$= \dfrac{32}{58}$

$= \dfrac{16}{29}$

3 a B = spaghetti bolognese; C = chilli con carne; V = vegetable curry

$n(B \cap C \cap V') = n(B \cap C) - n(B \cap C \cap V)$
$= 35 - 12$
$= 23$

$n(B \cap V \cap C') = n(B \cap V) - n(B \cap C \cap V)$
$= 20 - 12$
$= 8$

$n(B' \cap C \cap V) = n(C \cap V) - n(B \cap C \cap V)$
$= 24 - 12$
$= 12$

$n(B \cap C' \cap V') = n(B) - n(B \cap C) - n(B \cap V \cap C')$
$= 43 - 35 - 8$
$= 0$

$n(V \cap B' \cap C') = n(V) - n(V \cap B) - n(V \cap C \cap B')$
$= 80 - 20 - 12$
$= 48$

$n(C \cap V' \cap B') = 145 - 10 - 12 - 8 - 23 - 12 - 0 - 48 = 32$

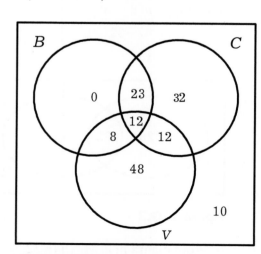

Figure 17G.3

b From diagram: $n(B \cap C' \cap V') = 0$

c $n(C) = 32 + 23 + 12 + 12 = 79$

d $P(1 \text{ meal only}) = \dfrac{0 + 32 + 48}{145}$

$= \dfrac{80}{145}$

$= \dfrac{16}{29}$

e $P(V | 1 \text{ meal only}) = \dfrac{n(V \text{ only})}{n(1 \text{ meal only})}$

$= \dfrac{48}{80}$

$= \dfrac{3}{5}$

f
$P(\text{at least 2 meals}) = 1 - P(0 \text{ meals}) - P(1 \text{ meal})$

$= 1 - \dfrac{n(0 \text{ meals})}{145} - \dfrac{3}{5}$

$= 1 - \dfrac{10}{145} - \dfrac{80}{145}$

$= \dfrac{11}{29}$

4 a B = blue eyes; D = dark hair

$P(B \cap D') = P(B) - P(B \cap D) = 0.4 - 0.2 = 0.2$

$P(B' \cap D) = P(D) - P(B \cap D) = 0.7 - 0.2 = 0.5$

$P(B' \cap D') = 1 - 0.2 - 0.2 - 0.5 = 0.1$

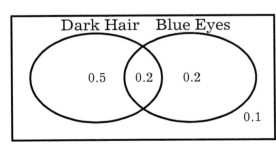

Figure 17G.4

b From diagram: $P(B' \cap D') = 0.1$

c $P(B|D) = \dfrac{P(B \cap D)}{P(D)}$

$= \dfrac{0.2}{0.7}$

$= \dfrac{2}{7}$

d $P(B|D') = \dfrac{P(B \cap D')}{P(D')}$

$= \dfrac{0.2}{1 - 0.7}$

$= \dfrac{2}{3}$

e From (c) and (d), $P(B|D) \neq P(B|D')$, so blue eyes and dark hair are not independent characteristics.

5 C = cold; R = raining
$P(C \cap R) = P(C) + P(R) - P(C \cup R)$

$= P(C) + P(R) - [1 - P((C \cup R)')]$

$= 0.6 + 0.45 - (1 - 0.25)$

$= 0.3$

b $P(C \cap R') = P(C) - P(C \cap R)$

$= 0.6 - 0.3$

$= 0.3$

$P(R \cap C') = P(R) - P(C \cap R)$

$= 0.45 - 0.3$

$= 0.15$

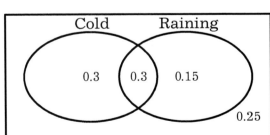

Figure 17G.5

c $P(C'|R) = \dfrac{P(R \cap C')}{P(R)} = \dfrac{0.15}{0.45} = \dfrac{1}{3}$

d $P(R|C') = \dfrac{P(R \cap C')}{P(C')}$

   $= \dfrac{0.15}{1-0.6}$

   $= \dfrac{3}{8}$

e $P(R \cap C) = 0.3$

   $P(R) \times P(C) = 0.6 \times 0.45 = 0.27$

   $P(R \cap C) \neq P(R) \times P(C)$, so the two events are not independent.

Exercise 17H

> **COMMENT**
>
> For the questions in this section, a tree diagram, appropriately populated with values, can be used to justify answers without further working.

1 Selection without replacement:

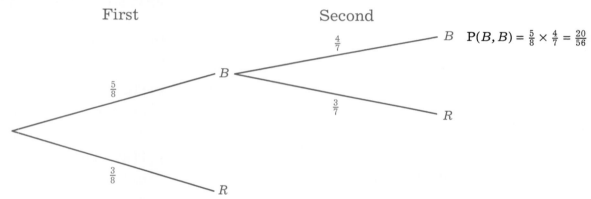

Figure 17H.1

B = blue; R = red

$P(B, B) = \dfrac{5}{8} \times \dfrac{4}{7}$

$= \dfrac{20}{56}$

$= \dfrac{5}{14}$

2 Selection without replacement:

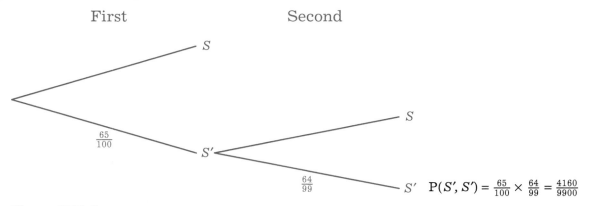

Figure 17H.2

S = strawberry flavoured

$P(\text{at least one } S) = 1 - P(S', S')$

$$= 1 - \frac{65}{100} \times \frac{64}{99}$$

$$= 1 - \frac{208}{495}$$

$$= \frac{287}{495}$$

$$= 0.580$$

3 Selection with replacement:

$$P(\text{blue, blue}) = P(\text{blue})^2 = \left(\frac{3}{5}\right)^2 = \frac{9}{25}$$

4 Selection without replacement:

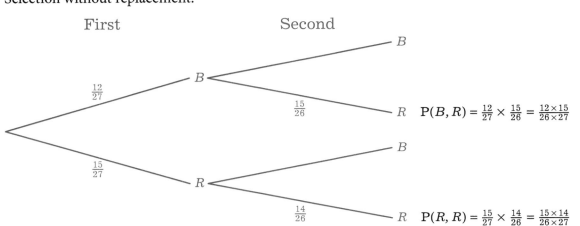

Figure 17H.4

B = black; R = red

$$P(R2) = P(R2|R1)P(R1) + P(R2|B1)P(B1)$$
$$= \frac{14}{26} \times \frac{15}{27} + \frac{15}{26} \times \frac{12}{27}$$
$$= \frac{210}{702} + \frac{180}{702}$$
$$= \frac{390}{702}$$
$$= \frac{5}{9}$$

COMMENT

Although this appears to be a problem involving conditional probability, this is an illusion. Since the question asks only about the second counter selection, without considering the result of the first selection, the probabilities for the second counter are the same as they would be if this were the first selection. The working above does not use this shortcut, but you should realise that this is the reason for the answer $\left(\frac{5}{9}\right)$ being the same as for drawing a single red counter $\left(\frac{15}{27}\right)$.

5 B = blue; O = orange

For selection with replacement:
$$P(\text{different}) = P(B,O) + P(O,B)$$
$$= \frac{1}{2} \times \frac{1}{2} + \frac{1}{2} \times \frac{1}{2}$$
$$= \frac{1}{4} + \frac{1}{4}$$
$$= \frac{1}{2}$$

For selection without replacement:

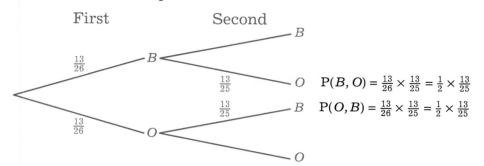

Figure 17H.5

$$P(B,O)+P(O,B)=\frac{13}{26}\times\frac{13}{25}+\frac{13}{26}\times\frac{13}{25}$$

$$=\frac{13}{25}>\frac{1}{2}$$

So a win is more likely if the tokens are selected without replacement.

6 a R = red; Y = yellow

For selection with replacement:

$$P(\text{different})=P(R,Y)+P(Y,R)$$

$$=\frac{1}{2}\times\frac{1}{2}+\frac{1}{2}\times\frac{1}{2}$$

$$=\frac{1}{4}+\frac{1}{4}$$

$$=\frac{1}{2}$$

b Let the number of each colour be n.

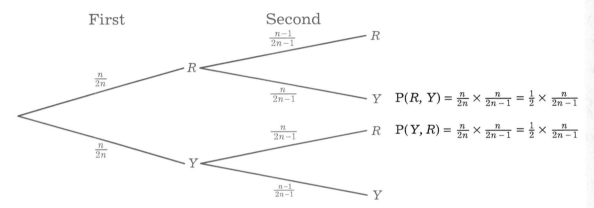

Figure 17H.6

$$P(R,Y)+P(Y,R)=\frac{n}{2n}\times\frac{n}{2n-1}+\frac{n}{2n}\times\frac{n}{2n-1}$$

$$=\frac{n}{2n-1}$$

Require that $\dfrac{n}{2n-1}<0.53$

$n<1.06n-0.53$

$0.06n>0.53$

$\Rightarrow n>88.3$

The least value of n which satisfies the condition is 9

$\therefore$ the smallest total number of balls satisfying the condition is $2\times 9=18$

Mixed examination practice 17

Short questions

1 $P(\text{same colour}) = P(R,R) + P(B,B) + P(W,W)$

$$= \frac{6}{18} \times \frac{5}{17} + \frac{4}{18} \times \frac{3}{17} + \frac{8}{18} \times \frac{7}{17}$$

$$= \frac{98}{306}$$

$$= \frac{49}{153}$$

$$= 0.320 \text{ (3SF)}$$

2 S = first language Spanish; A = Argentine

$$P(S|A) = \frac{P(S \cap A)}{P(A)}$$

$$= \frac{n(S \cap A)}{n(A)}$$

$$= \frac{12}{12+3}$$

$$= \frac{4}{5}$$

3

> **COMMENT**
> A tree diagram is the clearest and fastest way to show working in this case, but the full algebraic working is also given.

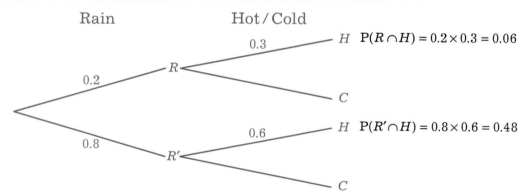

Figure 17MS.3

R = rain; H = hot (> 25°C); C = cold (≤ 25°C)

$$P(R|H) = \frac{P(R \cap H)}{P(H)} = \frac{0.2 \times 0.3}{0.2 \times 0.3 + 0.8 \times 0.6} = \frac{0.06}{0.54} = \frac{1}{9}$$

4 $P(A'|B) = \dfrac{P(A' \cap B)}{P(B)} = \dfrac{P(A') - P(A' \cap B')}{P(B)}$

$P(A') = 1 - P(A)$

$ = 1 - \dfrac{1}{14} = \dfrac{1}{15}$

$P(A' \cap B') = P\left((A \cup B)'\right) = 0$

$\therefore \dfrac{1}{5} = \dfrac{\dfrac{1}{15} - 0}{P(B)}$

$\Rightarrow P(B) = \dfrac{1}{3}$

Long questions

1 a R = red sweet

$P(R, R | \text{Large}) = \dfrac{8}{20} \times \dfrac{7}{19} = \dfrac{14}{95}$

b $P(R, R | \text{Small}) = \dfrac{4}{4+n} \times \dfrac{3}{3+n}$

$\therefore \dfrac{12}{(4+n)(3+n)} = \dfrac{2}{15}$

$90 = (4+n)(3+n)$

$n^2 + 7n - 78 = 0$

$(n-6)(n+13) = 0$

$\therefore n = 6$ (as $n = -13$ is not a valid solution in this context)

c $P(\text{Large}) = \dfrac{1}{3}$ and $P(\text{Small}) = \dfrac{2}{3}$

From (a), $P(R, R | \text{Large}) = \dfrac{14}{95}$

From (b), $P(R, R | \text{Small}) = \dfrac{4}{10} \times \dfrac{3}{9} = \dfrac{2}{15}$

$P(R, R) = P(R, R | \text{Large}) \times P(\text{Large}) + P(R, R | \text{Small}) \times P(\text{Small})$

$ = \dfrac{14}{95} \times \dfrac{1}{3} + \dfrac{2}{15} \times \dfrac{2}{3}$

$ = \dfrac{118}{855}$

$ = 0.138$

d $P(\text{Large}|R,R) = \dfrac{P(\text{Large} \cap \{R,R\})}{P(R,R)}$

$= \dfrac{\frac{42}{855}}{\frac{118}{855}}$

$= \dfrac{42}{118}$

$= \dfrac{21}{59}$

$= 0.356$

2 a Range of $P(X)$ is $[0, 1]$

b $P(A) - P(A \cap B) = P(A) - P(B|A)P(A)$
$= P(A)(1 - P(B|A))$

c i $P(A \cup B) = P(A) + P(B) - P(A \cap B)$
Hence $P(A \cup B) - P(A \cap B) = P(A) - P(A \cap B) + P(B) - P(A \cap B)$
From (b), $P(A) - P(A \cap B) = P(A)(1 - P(B|A))$
and similarly, $P(B) - P(A \cap B) = P(B)(1 - P(A|B))$
$\therefore P(A \cup B) - P(A \cap B) = P(A)(1 - P(B|A)) + P(B)(1 - P(A|B))$

ii The right-hand side of the equation in (i) is the sum of two products of non-negative values,
so $P(A \cup B) - P(A \cap B) \geq 0$
and hence $P(A \cup B) \geq P(A \cap B)$

3 a B = badminton; F = football
$P(B \cap F') = P(B) - P(B \cap F)$
$= 0.3 - x$
$P(F \cap B') = 1 - P(B) - P(F' \cap B')$
$= 1 - 0.3 - 0.5$
$= 0.2$

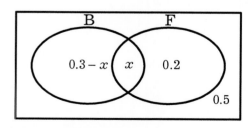

Figure 17ML.3.1

b $P(F \cap B') = 0.2$

c $P(B|F) = \dfrac{P(B \cap F)}{P(F)}$

$\therefore 0.5 = \dfrac{x}{0.2 + x}$

$0.5x + 0.1 = x$

$\Rightarrow x = P(B \cap F) = 0.2$

d

[Venn diagram: Two overlapping circles B and F in a rectangle. B-only region: 0.1; intersection: 0.2; F-only region: 0.2; outside: 0.5]

Figure 17ML.3.2

$P(B \cap F') = 0.1$

e $P(B \text{ only} | \text{one sport}) = \dfrac{P(B \text{ only})}{P(\text{one sport})}$

$= \dfrac{0.1}{0.1 + 0.2}$

$= \dfrac{1}{3}$

4 a i Since Ann takes the first throw,

$P(\text{Bridget wins on her first throw}) = P(6', 6)$

$= \dfrac{5}{6} \times \dfrac{1}{6}$

$= \dfrac{5}{36}$

ii $P(\text{Ann wins on her second throw}) = P(6', 6', 6)$

$= \dfrac{5}{6} \times \dfrac{5}{6} \times \dfrac{1}{6}$

$= \dfrac{25}{216}$

iii $P(\text{Ann wins on her } n\text{th throw})$

$= P(n-1 \text{ repeats of } \{6', 6'\} \text{ then followed by } 6)$

$= \left(\dfrac{5}{6} \times \dfrac{5}{6}\right)^{n-1} \times \dfrac{1}{6}$

$= \dfrac{5^{2n-2}}{6^{2n-1}}$

b If Ann and Bridget both fail to win on their first throws, the game effectively just starts again. That is,

P(Ann wins) = P(Ann wins on first throw) + P(6′, 6′) × P(Ann wins)

$$\therefore p = \frac{1}{6} + \frac{5}{6} \times \frac{5}{6} \times p$$

$$\Rightarrow p = \frac{1}{6} + \frac{25}{36}p$$

c $p = \frac{1}{6} + \frac{25}{36}p$

$$\frac{11}{36}p = \frac{1}{6}$$

$$\Rightarrow p = \frac{6}{11}$$

$$\therefore P(\text{Bridget wins}) = 1 - p = \frac{5}{11}$$

d

> **COMMENT**
>
> This part of the question requires knowledge of the binomial distribution from Section 18C.

Let X be the number of games won by Ann.

$$X \sim B\left(6, \frac{6}{11}\right)$$

$P(X > 3) = 0.432$ (from GDC)

18 Probability distributions

Exercise 18A

3 The sum of the probabilities must equal 1

$$\therefore \frac{1}{3}+\frac{1}{4}+k+\frac{1}{5}=1$$

$$\Rightarrow k=\frac{13}{60}$$

After two rolls the total score is 4. The cases are:

$$P(1,3)=\frac{1}{3}\times k=\frac{13}{180}$$

$$P(2,2)=\frac{1}{4}\times\frac{1}{4}=\frac{1}{16}$$

$$P(3,1)=k\times\frac{1}{3}=\frac{13}{180}$$

$$\therefore P(X_1+X_2=4)=\frac{13}{180}+\frac{1}{16}+\frac{13}{180}$$

$$=\frac{149}{720}$$

$$=0.207$$

Exercise 18B

2 a $\sum_x P(X=x)=1$

$k(3+4+5+6+7)=1$

$\Rightarrow k=\dfrac{1}{25}=0.04$

b $E(X)=\sum_x x\, P(X=x)$

$=k(2\times 3+3\times 4+4\times 5+5\times 6+6\times 7)$

$=\dfrac{110}{25}$

$=4.4$

3 $E(V)=\sum_x v\, P(V=v)=6.3$

$1\times 0.2+2\times 0.3+5\times 0.1+8\times 0.1+0.3k=6.3$

$2.1+0.3k=6.3$

$\Rightarrow k=\dfrac{4.2}{0.3}=14$

4 a $\sum_x P(X=x)=1$

$k(3+4+5+6)=1$

$\Rightarrow k=\dfrac{1}{18}$

b $E(X)=\sum_x x\, P(X=x)$

$=0\times 3k+1\times 4k+2\times 5k+3\times 6k$

$=32k$

$=\dfrac{16}{9}$

5 a $\sum_x P(X=x)=1$

$3k+4k+3k=1$

$\Rightarrow k=\dfrac{1}{10}$

b $E(X) = \sum_x x\, P(X=x)$

$= 1 \times 3k + 2 \times 4k + 3 \times 3k$

$= 20k$

$= 2$

6 $E(X) = \sum_x x\, P(X=x)$

$= (1+1+2+2+2+5) \times \dfrac{1}{6}$

$= \dfrac{13}{6}$

7 $\sum_x P(X=x) = 1$

$0.1 + p + q + 0.2 = 1$

$p + q = 0.7$

$\Rightarrow p = 0.7 - q$

$E(X) = \sum_x x\, P(X=x) = 1.5$

$0 \times 0.1 + 1 \times p + 2q + 3 \times 0.2 = 1.5$

$\Rightarrow p + 2q = 0.9$

$\therefore 0.7 - q + 2q = 0.9$

$\Rightarrow q = 0.2$

and so $p = 0.5$

8 Let X be the number of counters the player receives on a roll of the die.

To get an expected profit of 3.25 counters per roll when the player pays 5 counters per roll, require $E(X) = \sum_x x\, P(X=x) = 8.25$

$4 \times \dfrac{1}{2} + 5 \times \dfrac{1}{4} + 15 \times \dfrac{1}{5} + n \times \dfrac{1}{20} = \dfrac{33}{4}$

$\dfrac{n}{20} = 2$

$n = 40$

9 a Let a be the number that the player chooses.

The player's profit is equal to his winnings minus n.

$P(\text{Profit} = 3n) = P(a,a,a)$

$= \left(\dfrac{1}{4}\right)^3$

$= \dfrac{1}{64}$

$P(\text{Profit} = 2n)$
$= P(a,a,a') + P(a,a',a) + P(a',a,a)$

$= 3 \times \left(\dfrac{3}{4}\right) \times \left(\dfrac{1}{4}\right)^2$

$= \dfrac{9}{64}$

$P(\text{Profit} = 1 - n)$
$= P(a,a',a') + P(a',a,a') + P(a',a',a)$

$= 3 \times \left(\dfrac{3}{4}\right)^2 \times \left(\dfrac{1}{4}\right)$

$= \dfrac{27}{64}$

$P(\text{Profit} = n) = 1 - \dfrac{1}{64} - \dfrac{9}{64} - \dfrac{27}{64} = \dfrac{27}{64}$

TABLE 18B.9

Player's profit ($)	$-n$	$1-n$	$2n$	$3n$
Probability	$\dfrac{27}{64}$	$\dfrac{27}{64}$	$\dfrac{9}{64}$	$\dfrac{1}{64}$

b Let X be the player's profit. From the table in (a),

$E(X) = \sum_x x\, P(X=x)$

$= \dfrac{1}{64}(-27n + 27 - 27n + 18n + 3n)$

$= \dfrac{27 - 33n}{64}$

For the organiser to make a profit, require the player's profit to be negative:

$E(X) < 0$

$27 - 33n < 0$

$\Rightarrow n > 0.818$

∴ the minimum entrance fee is 82 cents.

Exercise 18C

5 a In a true binomial distribution, the probability of each 'success' is independent of the other results, as in the case of throwing a die multiple times. However, within the school population there is a fixed number of students who travel by bus, so the experiment is equivalent to drawing counters without replacement: each student chosen for the sample has a probability of bus travel which does depend on the previous students selected. Nevertheless, because the total number of students in the school is large and the base probability of 15% is not too close to 0% or 100%, the probabilities will not change very much, and so a binomial distribution will be a good approximation.

b Let X be the number of students in the sample travelling by bus.

We approximate $X \sim B(20, 0.15)$

$P(X = 5) = 0.103$ (from GDC)

6 $X \sim B\left(4050, \dfrac{2}{3}\right)$

$\Rightarrow E(X) = 4050 \times \dfrac{2}{3} = 2700$

7 Let X be the number of questions Sheila answers correctly.

$X \sim B\left(8, \dfrac{1}{4}\right)$

a From GDC: $P(X = 5) = 0.0231$

b $E(X) = 8 \times \dfrac{1}{4} = 2$

c $\text{Var}(X) = 8 \times \dfrac{1}{4} \times \dfrac{3}{4} = 1.5$

d From GDC: $P(X \geq 5) = 0.0273$

8 Let X be the number of people the doctor sees who have the virus.

$X \sim B(80, 0.008)$

a From GDC: $P(X = 2) = 0.108$

b From GDC: $P(X \geq 3) = 0.0267$

c It is assumed that patients do or do not have the virus independently of each other; in reality, if the prevalence nationwide is 0.8%, it is likely that there will be geographical pockets which have higher and lower rates than this, since a cold virus is contagious. Therefore, if the doctor sees one patient with the virus, it may be supposed that the virus is prevalent in the locality, so the probability of seeing other patients with the virus will be higher than 0.8%.

> **COMMENT**
>
> It is also assumed that the country is large enough that the doctor will not be seeing a significant fraction of the whole population; an island nation of a few hundred, for example, would not allow for a binomial model to be used with a sample of 80, for the same reasons as outlined in Q5(a). However, given the context of the question, this would not be the answer the examiner would be looking for!

9 $Y \sim B(12, p)$

$\Rightarrow E(Y) = 12p = 4.8$

$\Rightarrow p = 0.4$

10 Let X be the number of sixes in 4 throws: $X \sim B\left(4, \dfrac{1}{6}\right)$

$P(X = 3) = 0.0154$

Let Y be the number of fives or sixes in 6 throws: $Y \sim B\left(6, \dfrac{1}{3}\right)$

$P(Y = 5) = 0.0165$

So rolling 5 fives or sixes in 6 throws is the more probable event.

11 a $P(\text{Red 2nd} \mid \text{Blue 1st}) = \dfrac{5}{9}$

$P(\text{Red 2nd} \mid \text{Red 1st}) = \dfrac{4}{9}$

So first and second sock selections are not independent.

b $P(\text{Red 2nd})$

$= P(\text{Red 2nd} \mid \text{Blue 1st})P(\text{Blue 1st})$

$\quad + P(\text{Red 2nd} \mid \text{Red 1st})P(\text{Red 1st})$

$= \dfrac{5}{9} \times \dfrac{5}{10} + \dfrac{4}{9} \times \dfrac{5}{10}$

$= \dfrac{1}{2}$

$= P(\text{Red 1st})$

> **COMMENT**
>
> This is expected since the result of the first selection is not taken into consideration and is therefore irrelevant to the calculation; the scenario of drawing a sock, not looking at it and then checking the second sock drawn is equivalent to just drawing one sock and checking it. A similar result was seen in Exercise 17H question 4.

12 $X \sim B(n, 0.4)$

a $P(X = 2) = \binom{n}{2}(0.4)^2 (0.6)^{n-2}$

$= \dfrac{n(n-1)}{2} \times \dfrac{2^2}{5^2} \times \dfrac{3^{n-2}}{5^{n-2}}$

$= \dfrac{n(n-1)}{2} \times \dfrac{2^2 \times 3^{n-2}}{5^n}$

$= \dfrac{2n(n-1) \times 3^{n-2}}{5^n}$

b $\dfrac{2n(n-1) \times 3^{n-2}}{5^n} = 0.121$

$\Rightarrow n = 10$ (from GDC)

13 Let X be the number of sixes rolled in 12 throws.

$X \sim B(12, p)$

$P(X = 2) = \binom{12}{2} p^2 (1-p)^{10}$

$= 0.283$

From GDC: $p = 0.14$ or 0.20 (2DP)

14 $X \sim B(5, p)$

$P(X = 3) = \binom{5}{3} p^3 (1-p)^2 = 0.3087$

From GDC: $p = 0.494$ or 0.700 (3SF)

Exercise 18D

4 Let X be the life of a battery of this brand; then

$X \sim N(16, 5^2)$, $x = 10.2$

a $\dfrac{\mu - x}{\sigma} = \dfrac{16 - 10.2}{5} = 1.16$ standard deviations below the mean

b $P(X < 10.2) = P(Z < -1.16)$

$= 0.123$ (from GDC)

5 Let X be the weight of a cat of this breed; then $X \sim N(16, 4^2)$

$P(X>13) = 0.773$ (from GDC)

$\therefore$ expected number in a sample of 2000 is
$0.773 \times 2000 = 1547$

6 Let X be the length of one of Ali's jumps; then $X \sim N(5.2, 0.7^2)$

a $P(5<X<5.5) = P(X<5.5) - P(X<5)$
$= 0.666 - 0.388$
$= 0.278$ (from GDC)

b i $P(X \geq 6) = 0.127$ (from GDC)

ii $P(\text{Qualify}) = 1 - P(\text{Fail three times})$
$= 1 - (P(X<6))^3$
$= 0.334$ (from GDC)

7 $D \sim N(250, 20^2)$

a $P(265<D<280) = P(D<280) - P(D<265)$
$= 0.933 - 0.773$
$= 0.160$ (from GDC)

b $P(D>265 \mid D<280) = \dfrac{P(265<D<280)}{P(D<280)}$
$= \dfrac{0.160}{0.933}$
$= 0.171$

c $P(D<242 \cup D>256)$
$= 1 - P(242 \leq D \leq 256)$
$= 1 - P(D \leq 256) + P(D<242)$
$= 1 - 0.618 + 0.345$
$= 0.727$

8 $Q \sim N(4, 160)$

a $P(|Q|>5) = 1 - P(-5 \leq Q \leq 5)$
$= 1 - P(Q \leq 5) + P(Q<-5)$
$= 1 - 0.532 + 0.238$
$= 0.707$ (from GDC)

b $P(Q>5 \mid |Q|>5) = \dfrac{P(Q>5 \cap |Q|>5)}{P(|Q|>5)}$
$= \dfrac{P(Q>5)}{P(|Q|>5)}$
$= \dfrac{1-0.532}{0.707}$
$= 0.663$ (from GDC)

9 Let X be the weight of an apple in this shipment; then
$X \sim N(150, 25^2)$

a
$P(120<X<170) = P(X<170) - P(X<120)$
$= 0.788 - 0.115$
$= 0.673$ (from GDC)

b Let Y be the number of medium apples in a bag of 10; then
$Y \sim B(10, 0.673)$
From GDC: $P(Y \geq 8) = 0.314$

10 Let X be the wingspan of a pigeon; then
$X \sim N(60, 6^2)$

a From GDC: $P(X>50) = 0.952$

b $P(X>55 \mid X>50) = \dfrac{P(X>55 \cap X>50)}{P(X>50)}$
$= \dfrac{P(X>55)}{P(X>50)}$
$= \dfrac{0.798}{0.952}$
$= 0.838$ (from GDC)

18 Probability distributions

11 Let X be the width of a grain of sand; then
$X \sim N(2, 0.5^2)$

  a From GDC: $P(X>1.5)=0.841$

  b
$$P(X>1.5 \mid X<2.5) = \frac{P(1.5<X<2.5)}{P(X<2.5)}$$
$$= \frac{P(X<2.5)-P(X<1.5)}{P(X<2.5)}$$
$$= \frac{0.841-0.159}{0.841}$$
$$= 0.811 \text{ (from GDC)}$$

12 Let X be the amount of paracetamol in a tablet; then
$X \sim N(500, 160^2)$
$P(X<300)=0.106$ (from GDC)

Let Y be the number of people in the sample of 20 who get a less than effective dose.
$Y \sim B(20, 0.106)$
$\Rightarrow P(Y \geq 2) = 0.640$ (from GDC)

Exercise 18E

4 Let X be the score in an IQ test; then
$X \sim N(100, 20^2)$

$P(X>x)=2\%$
$\Rightarrow P(X \leq x)=0.98$
$\Rightarrow x=141$ (from GDC)

5 Let X be the mass of a rabbit; then
$X \sim N(2.6, 1.2^2)$

$P(X \geq x)=20\%$
$\Rightarrow P(X<x)=0.8$
$\Rightarrow x=3.61$ kg (from GDC)

6 Let X be the diameter of a bolt; then
$X \sim N(\mu, 0.02^2)$

$P(X>2)=6\%$
$\Rightarrow P(X \leq 2)=0.94$
$\Rightarrow \frac{2-\mu}{0.02} = \Phi^{-1}(0.94)=1.55$ (from GDC)
$\Rightarrow \mu = 2-1.55 \times 0.02 = 1.97$ cm

7 Let X be the time a student takes to complete the test; then
$X \sim N(32, 6^2)$

  a From GDC: $P(X<35)=0.691$

  b $P(X<t)=0.9$
$\Rightarrow t=39.7$ minutes (39 minutes and 41 seconds)

  c $P(X<30)=0.369$
Let Y be the number of the 8 students who completed the test in less than 30 minutes.
$Y \sim B(8, 0.369)$
$\Rightarrow P(Y=2)=0.240$

8 If $X \sim N(\mu, \sigma^2)$, then (from GDC)
$P(\mu-3\sigma < X < \mu+3\sigma) = \Phi(3)-\Phi(-3)$
$= 0.99865-0.00135$
$= 0.9973$
$= 99.73\%$

> **COMMENT**
> This assumes that the range is centred on the mean, which, while not necessarily the case, is a valid approximation.

9 a Let G be the diameter of a grain of sand from Playa Gauss.

$G \sim N(\mu_G, \sigma_G^2)$

$P(G < 1) = 0.3$

$\Rightarrow \dfrac{1 - \mu_G}{\sigma_G} = \Phi^{-1}(0.3) = -0.524$

$\Rightarrow \mu_G = 1 + 0.524\sigma_G$ ... (1)

$P(G < 2) = 0.9$

$\Rightarrow \dfrac{2 - \mu_G}{\sigma_G} = \Phi^{-1}(0.9) = 1.28$

$\Rightarrow \mu_G = 2 - 1.28\sigma_G$ ... (2)

Substituting (2) into (1):

$2 - 1.28\sigma_G = 1 + 0.524\sigma_G$

$1.81\sigma_G = 1$

$\therefore \sigma_G = 0.554$ mm

and hence $\mu_G = 1.29$ mm

b Let F be the diameter of a grain of sand from Playa Fermat.

$F \sim N(\mu_F, \sigma_F^2)$

$P(F < 2) = 0.8$

$\Rightarrow \dfrac{2 - \mu_F}{\sigma_F} = \Phi^{-1}(0.8) = 0.842$

$\Rightarrow \mu_F = 2 - 0.842\sigma_F$ ... (1)

$P(F < 1) = 80\% \times 40\% = 0.32$

$\Rightarrow \dfrac{1 - \mu_F}{\sigma_F} = \Phi^{-1}(0.32) = -0.468$

$\Rightarrow \mu_F = 1 + 0.468\sigma_F$ ... (2)

Substituting (2) into (1):

$1 + 0.468\sigma_F = 2 - 0.842\sigma_F$

$1.31\sigma_F = 1$

$\therefore \sigma_F = 0.764$ mm

and hence $\mu_F = 1.36$ mm

Mixed examination practice 18
Short questions

> **COMMENT**
>
> It is always wise to define your random variable clearly at the start of a question, especially in situations such as Q2 where you may want to consider the loss or profit explicitly rather than just the outcome of a die roll, or Q5 where you will use one distribution result to inform a different distribution.
> Defining your variable at the start of working makes the calculations clear to the reader, whether that is the examiner or yourself, when checking your answer.

1 Let X be the number of defective bottles in a sample of 20.

$X \sim B(20, 0.015)$

$P(X \geq 1) = 1 - P(X = 0)$

$\qquad\qquad = 1 - 0.985^{20}$

$\qquad\qquad = 26.1\%$

2 Let X be the expected loss on each play; then

$X = 10 - (\text{value on die})$

$E(X) = \displaystyle\sum_x x\, P(X = x)$

$\qquad = 10 - \left(1 \times \dfrac{1}{2} + 5 \times \dfrac{1}{5} + 10 \times \dfrac{1}{5} + \dfrac{N}{10}\right)$

$\qquad = \dfrac{13}{2} - \dfrac{N}{10}$

Require $E(X) = 1$

$\dfrac{13}{2} - \dfrac{N}{10} = 1$

$\dfrac{N}{10} = \dfrac{11}{2}$

$\Rightarrow N = 55$

3 Let X be the test score of a student.
$X \sim N(62, 12^2)$

  a From GDC: $P(X > 80) = 0.0668 = 6.68\%$

  b $P(X \geq x) = 5\%$
  $\Rightarrow P(X < x) = 0.95$
  $\Rightarrow x = 81.7$ (from GDC)

  ∴ the lowest score achieved by a student in the top 5% is 81, assuming integer scores.

> **COMMENT**
> Note that 81.7 does not round up to 82, since fewer than 5% of students would score at least 82, and we are asked for the lowest score of the top 5%, which must therefore include some students who scored below 82.

4 $X \sim B(12, 0.4)$

  a $E(X) = 12 \times 0.4 = 4.8$ balls

  b $Var(X) = 12 \times 0.4 \times 0.6 = 2.88$

  $P(X \leq 2.88) = P(X \leq 2)$
  $= 0.0834$

5 Let X be the height of a dog of this breed; then
$X \sim N(0.7, 0.05)$

From GDC: $P(X > 0.75) = 0.412$

> **COMMENT**
> Remember to use the full value from your calculator in further calculation, rather than the 3SF value you may write in working.

Let Y be the number of dogs in a sample of 6 with height greater than 0.75 m; then
$Y \sim B(6, 0.412)$

> **COMMENT**
> Always use a different letter for each variable to keep your working clear.

From GDC: $P(Y = 4) = 0.149$

6 Let X be the number of times Robyn hits the target in 8 attempts; then
$X \sim B(8, 0.6)$

  a From GDC: $P(X = 4) = 0.232$

  b $P(\text{fails to qualify}) = P(X \leq 6)$
  $= 0.894$ (from GDC)

7 Let X be the number of defective bulbs in a pack of 6; then
$X \sim B(6, 0.005)$

  a $P(X \geq 1) = 1 - P(X = 0)$
  $= 1 - (0.995)^6$
  $= 0.0296$

  b Let Y be the number of packs in a sample of 20 that contain at least one defective bulb.
  $Y \sim B(20, 0.0296)$
  ∴ $P(Y > 4) = 0.000244 = 0.0244\%$
  (from GDC)

8 Let X be an estimate of the angle (in degrees).
$X \sim N(\mu, \sigma^2)$

$P(X < 25) = \dfrac{16}{200} = 0.08$

$\Rightarrow \dfrac{25 - \mu}{\sigma} = \Phi^{-1}(0.08) = -1.41$

$\Rightarrow \mu = 25 + 1.41\sigma$ ... (1)

$P(X > 35) = \dfrac{42}{200} = 0.21$

$\Rightarrow P(X \leq 35) = 0.79$

$\Rightarrow \dfrac{35 - \mu}{\sigma} = \Phi^{-1}(0.79) = 0.806$

$\Rightarrow \mu = 35 - 0.806\sigma \quad \ldots (2)$

Substituting (2) into (1):

$35 - 0.806\sigma = 25 + 1.41\sigma$

$2.21\sigma = 10$

$\Rightarrow \sigma = 4.52°$

and hence $\mu = 31.4°$

9 Let X be the number of sixes from n rolls; then

$X \sim B\left(n, \dfrac{1}{6}\right)$

a $P(X \leq 2) = P(X=0) + P(X=1) + P(X=2)$

$= \binom{n}{0}\left(\dfrac{5}{6}\right)^n + \binom{n}{1}\left(\dfrac{5}{6}\right)^{n-1}\left(\dfrac{1}{6}\right) + \binom{n}{2}\left(\dfrac{5}{6}\right)^{n-2}\left(\dfrac{1}{6}\right)^2$

$= \dfrac{5^{n-2}}{6^n}\left(5^2 + 5n + \dfrac{n(n-1)}{2}\right)$

$= \dfrac{5^{n-2}}{2 \times 6^n}\left(50 + 9n + n^2\right)$

Require that $\dfrac{5^{n-2}}{2 \times 6^n}\left(50 + 9n + n^2\right) = 0.532$

From GDC, $n = 15$

b $X \sim B\left(15, \dfrac{1}{6}\right)$

$\Rightarrow P(X = 2) = 0.273$

Long questions

1 Let Y be the number of yellow ribbons in the sample of 10.

$Y \sim B\left(10, \dfrac{1}{4}\right)$

a $E(Y) = 10 \times \dfrac{1}{4} = 2.5$

b $P(Y = 6) = 0.0162$ (from GDC)

c $P(Y \geq 2) = 0.756$ (from GDC)

d Expect the mode to be close to the mean for a binomial distribution.
   From GDC:
   $P(Y=1) = 18.8\%$
   $P(Y=2) = 28.2\%$
   $P(Y=3) = 25.0\%$
   $P(Y=4) = 14.6\%$
   From the above, the mode is 2.

e Have assumed that $P(\text{yellow}) = 0.25$ is constant.

> **COMMENT**
>
> In using a binomial distribution, we assume that each choice is independent of the previous one — that is, the probability of drawing yellow is the same each time. Since we are told that the bag contains a very large number of ribbons, this is approximately true — P(yellow) does not change much, because even after removing several ribbons, the proportion of the remaining ribbons which are yellow stays approximately one-quarter.

2 Let X be the number of students who forget to do homework.

a $X \sim B(12, 0.05)$
   $P(X \geq 1) = 0.460$

b $X \sim B(n, 0.05)$
   $P(X \geq 1) = 1 - P(X = 0)$
   $\qquad\qquad = 1 - 0.95^n$

c Require $1 - 0.95^n \geq 0.8$
   $0.95^n \leq 0.2$
   $n \geq \dfrac{\log 0.2}{\log 0.95}$
   $n \geq 31.4$

   The smallest number of students under this requirement is 32.

> **COMMENT**
>
> Remember that $\log 0.95 < 0$, so the inequality is reversed when dividing through by it.

3

TABLE 18ML.3.1

		Die 1					
	Score	1	2	3	4	5	6
Die 2	1	2	3	4	5	6	7
	2	3	4	5	6	7	8
	3	4	5	6	7	8	9
	4	5	6	7	8	9	10
	5	6	7	8	9	10	11
	6	7	8	9	10	11	12

Let A be Alan's score.

TABLE 18ML.3.2

a	2	3	4	5	6	7	8	9	10	11	12
$P(A=a)$	$\frac{1}{36}$	$\frac{2}{36}$	$\frac{3}{36}$	$\frac{4}{36}$	$\frac{5}{36}$	$\frac{6}{36}$	$\frac{5}{36}$	$\frac{4}{36}$	$\frac{3}{36}$	$\frac{2}{36}$	$\frac{1}{36}$

a i $P(A=9) = \frac{4}{36} = \frac{1}{9}$

   ii Let B be Belle's score.
   The scores of Alan and Belle are independent events with the same distribution.
   $$\therefore P(A=9 \cap B=9) = \left(\frac{1}{9}\right)^2 = \frac{1}{81}$$

b i $P(A=B) = \sum_{x}(P(X=x))^2$

   $= \frac{1}{36^2}(1^2 + 2^2 + 3^2 + 4^2 + 5^2 + 6^2 + 5^2 + 4^2 + 3^2 + 2^2 + 1^2)$

   $= \frac{146}{36^2}$

   $= \frac{73}{648}$

   $= 0.113$

ii By symmetry, $P(A > B) = P(B > A)$

$\therefore P(A > B) = \dfrac{1}{2}(1 - P(A = B))$

$= 0.444$

c i The number shown on each die has the same uniform distribution.

$P(X \leq x) = P(\text{Roll} \leq x \text{ four times})$

$= (P(\text{Roll} \leq x))^4$

$= \left(\dfrac{x}{6}\right)^4$

ii $P(X = x) = P(X \leq x) - P(X \leq x - 1)$

$= \dfrac{x^4}{6^4} - \dfrac{(x-1)^4}{6^4}$

$= \dfrac{x^4 - (x-1)^4}{6^4}$

TABLE 18ML.3.3

x	1	2	3	4	5	6
P(X = x)	$\dfrac{1}{1296}$	$\dfrac{15}{1296}$	$\dfrac{65}{1296}$	$\dfrac{175}{1296}$	$\dfrac{369}{1296}$	$\dfrac{671}{1296}$

iii $E(X) = \sum\limits_{x} x\, P(X = x)$

$= \dfrac{1}{1296}(1 \times 1 + 2 \times 15 + 3 \times 65 + 4 \times 175 + 5 \times 369 + 6 \times 671)$

$= \dfrac{6797}{1296}$

$= 5.24$

19 Questions crossing chapters

Short questions

1 The number of possible outcomes is 36.

An average of 3 is equivalent to a sum of 6, and the outcomes that give this result are

1+5
2+4
3+3
4+2
5+1

Each has probability $\frac{1}{36}$, so the total probability is $\frac{5}{36}$.

2 The first two terms are:

$u_1 = S_1$
$\quad = 3(1) + 2(1)^2$
$\quad = 5$

$u_2 = S_2 - S_1$
$\quad = 3(2) + 2(2)^2 - 5$
$\quad = 14 - 5$
$\quad = 9$

$\therefore d = u_2 - u_1 = 4$

> **COMMENT**
>
> We can also find d by comparing the given expression for S_n with the general formula
>
> $S_n = \frac{n}{2}(2u_1 + (n-1)d)$
>
> $\quad = \left(u_1 - \frac{1}{2}d\right)n + \frac{1}{2}dn^2$
>
> Comparing the coefficient of n^2 in the general formula with that in $S_n = 3n + 2n^2$, we get $\frac{1}{2}d = 2 \Rightarrow d = 4$.

3 $|x| = \begin{cases} x & \text{for } x \geq 0 \\ -x & \text{for } x < 0 \end{cases}$

The graph is composed of two straight lines, one for positive x with gradient 1 and one for negative x with gradient -1, which meet at the origin.

Therefore the graph of $f'(x)$ is as follows:

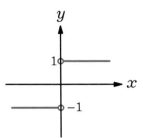

Figure 19S.3

4 $\boldsymbol{u}\cdot\boldsymbol{v} = x^2 + x + x$

$= x^2 + 2x$

$\therefore \int \boldsymbol{u}\cdot\boldsymbol{v}\,\mathrm{d}x = \int x^2 + 2x\,\mathrm{d}x$

$= \dfrac{1}{3}x^3 + x^2 + c$

5 Average $= \dfrac{S_n}{n}$

$= \dfrac{1}{n}\left(\dfrac{a(1-r^n)}{1-r}\right)$

$= \dfrac{a(1-r^n)}{n(1-r)}$

6 Mean $= \dfrac{3 + x^2 + x}{3}$

$= \dfrac{1}{3}x^2 + \dfrac{1}{3}x + 1$

This is a positive quadratic with vertex at

$\dfrac{-\dfrac{1}{3}}{2\left(\dfrac{1}{3}\right)} = -\dfrac{1}{2}$

Hence the minimum possible value is

$\dfrac{1}{3}\left(-\dfrac{1}{2}\right)^2 + \dfrac{1}{3}\left(-\dfrac{1}{2}\right) + 1 = \dfrac{1-2+12}{12} = \dfrac{11}{12}$

> **COMMENT**
>
> We could also use completing the square or differentiation to find the minimum, but it's worth remembering that the equation for the line of symmetry is given in the formula book.

7 $\sum\limits_x P(X = x) = 1$

$\ln k + \ln 2k + \ln 3k + \ln 4k = 1$

$\ln(24k^4) = 1$

$24k^4 = \mathrm{e}$

$k = \sqrt[4]{\dfrac{\mathrm{e}}{24}}$

8 The graph of $y = 1 - \cos x$ is obtained from the graph of $y = \cos x$ by reflection in the x-axis followed by translation one unit up:

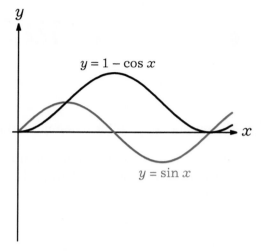

Figure 19S.8

Compared with $y = \sin x$, this graph is translated $\dfrac{\pi}{2}$ units to the right and 1 unit up, so the single transformation is the translation with vector $\begin{pmatrix}\dfrac{\pi}{2}\\1\end{pmatrix}$.

9 a The horizontal stretch scale factor is $\dfrac{1}{k}$.

b $y = \ln kx$

$= \ln k + \ln x$

So the vertical translation is by $\ln k$ units up.

10 a This is a geometric series with first term $0.5^0 = 1$ and common ratio 0.5. There are 11 terms in the sum. So

$S_{11} = \dfrac{1(1 - 0.5^{11})}{1 - 0.5}$

$= \dfrac{1 - \dfrac{1}{2^{11}}}{\dfrac{1}{2}}$

$= \dfrac{2^{11} - 1}{2^{10}}$

$= \dfrac{2047}{1024}$

b $\ln u_r = \ln 0.5^r$
$= r \ln 0.5$

$\sum_{r=0}^{10} \ln(u_r) = \ln 0.5 \sum_{r=0}^{10} r$

$= \ln 0.5 (0 + 1 + \cdots + 10)$

$= 55 \ln 0.5$

$= 55 \ln 2^{-1}$

$= -55 \ln 2$

11 $f(g(x)) = 3(ax^2 - x + 5) + 1$

$= 3ax^2 - 3x + 16$

$f(g(x)) = 0$ is a quadratic equation and has equal roots when $\Delta = 0$:

$(-3)^2 - 4 \times 3a \times 16 = 0$

$9 - 192a = 0$

$a = \dfrac{3}{64}$

12 $\int_0^y x^2 + 1 \, dx = 4$

$\left[\dfrac{1}{3} x^3 + x \right]_0^y = 4$

$\dfrac{1}{3} y^3 + y - 4 = 0$

From GDC: $y = 1.86$

13 This is a geometric series with common ratio $r = x^2 - x$. It converges when $|x^2 - x| < 1$.

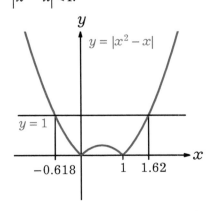

Figure 19S.13 Graphs of $y = |x^2 - x|$ and $y = 1$

From the figure, $|x^2 - x| < 1$ for $-0.618 < x < 1.62$.

14 $\mathbf{u} \cdot \mathbf{v} = 0 \Rightarrow \mathbf{u}$ and $\mathbf{v}$ are perpendicular.

The vector $\mathbf{u} - \mathbf{v}$ is the diagonal of the rectangle, as shown in Figure 19S.14.

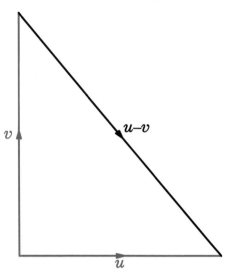

Figure 19S.14

$|\mathbf{u} - \mathbf{v}|^2 = \sqrt{|\mathbf{u}|^2 + |\mathbf{v}|^2} = \sqrt{2^2 + 3^2} = \sqrt{13}$

COMMENT

We can also solve the problem algebraically, using $|\mathbf{v}|^2 = \mathbf{v} \cdot \mathbf{v}$:

$|\mathbf{u} - \mathbf{v}|^2 = (\mathbf{u} - \mathbf{v}) \cdot (\mathbf{u} - \mathbf{v})$

$= |\mathbf{u}|^2 + |\mathbf{v}|^2 - 2\mathbf{u} \cdot \mathbf{v}$

$= 2^2 + 3^2 - 2 \times 0$

$= 13$

$\therefore |\mathbf{u} - \mathbf{v}| = \sqrt{13}$

15 a

$P(\text{first 6 on third roll}) = P(6') \times P(6') \times P(6)$

$= \dfrac{5}{6} \times \dfrac{5}{6} \times \dfrac{1}{6}$

$= \dfrac{25}{216}$

b To get the first six on the rth roll, he rolls $r-1$ non-sixes followed by a six:

$$p_r = \left(\frac{5}{6}\right)^{r-1}\left(\frac{1}{6}\right) = \frac{5^{r-1}}{6^r}$$

c $\sum_{r=1}^{\infty} p_r = \sum_{r=1}^{\infty}\left(\frac{1}{6}\right)\left(\frac{5}{6}\right)^{r-1}$

This is a geometric series with first term $\frac{1}{6}$ and common ratio $\frac{5}{6}$, so

$$S_\infty = \frac{\frac{1}{6}}{1-\frac{5}{6}}$$

$$= \frac{1/6}{1/6}$$

$$= 1$$

16 The sum is a geometric series with first term 1 and common ratio x (which converges because $0 < x < 1$), so $\sum_{r=1}^{\infty} x^r = \frac{1}{1-x}$.

Hence $\int_0^{1/2} \sum_{r=1}^{\infty} x^r \, dx = \int_0^{1/2} \frac{1}{1-x} \, dx$

$$= \left[-\ln(1-x)\right]_0^{1/2}$$

$$= -\ln\left(\frac{1}{2}\right) + \ln(1)$$

$$= \ln 2$$

17 Need to express the volume V of the cone in terms of θ.

From the diagram, the height is $h = l\cos\theta$ and the radius of the base is $l\sin\theta$.

$$\therefore V = \frac{1}{3}\pi(l\sin\theta)^2(l\cos\theta)$$

$$= \frac{1}{3}\pi l^3 \sin^2\theta \cos\theta$$

From GDC, the maximum value of $y = \sin^2\theta\cos\theta$ for $\theta \in \left(0, \frac{\pi}{2}\right)$ is 0.385, so the maximum possible volume is

$$V_{max} = \frac{1}{3} \times 0.385\pi l^3 = 0.403 l^3$$

COMMENT

Although the maximum value can be found directly using a GDC, differentiation could also be used: local maximum of V occurs where $\dfrac{dV}{d\theta} = 0$

$$\therefore \frac{1}{3}\pi l^3 \left(2\sin\theta \cos^2\theta - \sin^3\theta\right) = 0$$

$$\sin\theta \left(2\cos^2\theta - \sin^2\theta\right) = 0$$

$$\sin\theta = 0 \quad \text{or} \quad 2\cos^2\theta - \sin^2\theta = 0$$

$$\therefore 2\cos^2\theta - \sin^2\theta = 0 \quad \left(\text{as } \sin\theta \ne 0 \text{ for } 0 < \theta < \frac{\pi}{2}\right)$$

$$\frac{\sin^2\theta}{\cos^2\theta} = 2$$

$$\tan^2\theta = 2$$

$$\therefore \tan\theta = \sqrt{2} \quad \left(\tan\theta > 0 \text{ for } 0 < \theta < \frac{\pi}{2}\right)$$

Hence, using a $1, \sqrt{2}, \sqrt{3}$ right-angled triangle, $\cos\theta = \dfrac{1}{\sqrt{3}}$.

At this value of θ,

$$V = \frac{1}{3}\pi l^3 \tan^2\theta \cos^3\theta$$

$$= \frac{1}{3}\pi l^3 \times 2 \times \frac{1}{3\sqrt{3}}$$

$$= \frac{2\pi l^3}{9\sqrt{3}} = 0.403 l^3$$

18 a ϕ is an interior angle of a parallelogram, so $\phi = \pi - 2\theta$

b $\boldsymbol{a} + \boldsymbol{b}$ is the longer diagonal of the parallelogram.
Using the cosine rule in the triangle with sides $|\boldsymbol{a}|$ and $|\boldsymbol{b}|$ and angle ϕ between them,

$$|\boldsymbol{a}+\boldsymbol{b}|^2 = |\boldsymbol{a}|^2 + |\boldsymbol{b}|^2 - 2|\boldsymbol{a}||\boldsymbol{b}|\cos\phi$$

$$= 1 + 1 - 2\cos(\pi - 2\theta)$$

$$= 2 - 2(-\cos 2\theta)$$

$$= 2 + 2(2\cos^2\theta - 1)$$

$$= 4\cos^2\theta$$

19 a $f(x) = \ln(x^2 - 9) - \ln(x+3) - \ln x$

$= \ln\left(\dfrac{x^2 - 9}{x(x+3)}\right)$

$= \ln\left(\dfrac{(x-3)(x+3)}{x(x+3)}\right)$

$= \ln\left(\dfrac{x-3}{x}\right)$

> **COMMENT**
>
> It is possible to cancel the fraction since $x \neq -3$ for the function to be defined.

b $y = \ln\left(\dfrac{x-3}{x}\right)$

$\dfrac{x-3}{x} = e^y$

$xe^y - x = -3$

$x(e^y - 1) = -3$

$\Rightarrow x = \dfrac{-3}{e^y - 1}$

$\therefore f^{-1}(x) = \dfrac{3}{1 - e^x}$

20 a $\ln(x^2) = 2\ln x$, so the transformation is a vertical stretch with scale factor 2.

b $\log_{10} x = \dfrac{\ln x}{\ln 10}$, so the transformation is a vertical stretch with scale factor $\dfrac{1}{\ln 10}$.

21 $P(X = x) = \dfrac{4p^x}{5}$ is a geometric sequence with first term $\dfrac{4(p)^0}{5} = \dfrac{4}{5}$ and common ratio p.

$\sum_{x \in \mathbb{N}} P(X = x)$ is an infinite geometric series.

Using the formula for S_∞,

$\sum_x P(X = x) = 1$

$\dfrac{\frac{4}{5}}{1 - p} = 1$

$1 - p = \dfrac{4}{5}$

$\therefore p = \dfrac{1}{5}$

22 a Setting $x = 1$ in $(1+x)^n$ gives

$(1+1)^n = \sum_{r=0}^{n} \binom{n}{r} 1^{n-r} 1^r$

$\sum_{r=0}^{n} \binom{n}{r} = 2^n$

b Setting $x = -1$ in $(1+x)^n$:

$(1-1)^n = \sum_{r=0}^{n} \binom{n}{r} 1^{n-r} (-1)^r$

$\Rightarrow \sum_{r=0}^{n} (-1)^r \binom{n}{r} = 0$

23 The length of an arc is $l = r\theta$, so need to find the radii OB, OB_1, OB_2 etc.

$OA = OB = 1$

$OA_1 = OB_1$
$= OA \cos\theta$
$= 1 \times \cos\theta$
$= \cos\theta$

$OA_2 = OB_2$
$= OA_1 \cos\theta$
$= \cos\theta \times \cos\theta$
$= \cos^2\theta$

Hence the radii form a geometric sequence with first term 1 and common ratio $\cos\theta$.

$AB + A_1B_1 + A_2B_2 + \cdots$
$= 1 \times \theta + (\cos\theta)\theta + (\cos^2\theta)\theta + \cdots$
$= (1 + \cos\theta + \cos^2\theta + \cdots)\theta$
$= \left(\dfrac{1}{1-\cos\theta}\right)\theta$ (using formula for S_∞)
$= \dfrac{\theta}{1-\cos\theta}$

24 a $y = e^x + \dfrac{1}{e^x}$

$\Rightarrow e^{2x} - ye^x + 1 = 0$

This is a quadratic equation in e^x; using the quadratic formula,

$e^x = \dfrac{y \pm \sqrt{y^2 - 4}}{2}$

$\Rightarrow x = \ln\left(\dfrac{y \pm \sqrt{y^2 - 4}}{2}\right)$

b Summing these two roots,

$x_1 + x_2 = \ln\left(\dfrac{y + \sqrt{y^2 - 4}}{2}\right) + \ln\left(\dfrac{y - \sqrt{y^2 - 4}}{2}\right)$

$= \ln\left(\dfrac{(y+\sqrt{y^2-4})(y-\sqrt{y^2-4})}{4}\right)$

$= \ln\left(\dfrac{y^2 - (y^2 - 4)}{4}\right)$

$= \ln\left(\dfrac{4}{4}\right)$

$= 0$

25

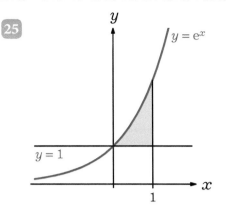

Figure 19S.25

This is the same volume as when the region between the graph of $y = e^x - 1$, the x-axis and the line $x = 1$ is rotated around the x-axis. (The whole picture is just translated vertically by one unit.)

$\therefore V = \pi \int_0^1 (e^x - 1)^2 \, dx$

$= \pi \int_0^1 e^{2x} - 2e^x + 1 \, dx$

$= \pi \left[\dfrac{1}{2}e^{2x} - 2e^x + x\right]_0^1$

$= \pi\left[\left(\dfrac{1}{2}e^2 - 2e + 1\right) - \left(\dfrac{1}{2}e^0 - 2e^0 + 0\right)\right]$

$= \pi\left(\dfrac{1}{2}e^2 - 2e + \dfrac{5}{2}\right)$

26 Probability has to be between 0 and 1, so need all integers x for which

$0 \leq \dfrac{1}{7}(x^2 - 14x + 38) \leq 1$

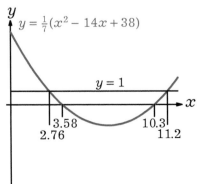

Figure 19S.26

19 Questions crossing chapters 221

From the graph on the GDC, the inequality is satisfied for $2.76 \leq x \leq 3.68$ and $10.3 \leq x \leq 11.2$.

Hence the possible integer values of x are 3 and 11.

27 Using $\mathbf{a} \cdot \mathbf{a} = |\mathbf{a}|^2$, $\mathbf{b} \cdot \mathbf{b} = |\mathbf{b}|^2$ and $|\mathbf{a}| = |\mathbf{b}| = x$:

$(\mathbf{a}+\mathbf{b}) \cdot (\mathbf{a}+\mathbf{b}) = 6x$
$\Rightarrow \mathbf{a} \cdot \mathbf{a} + \mathbf{b} \cdot \mathbf{b} + 2\mathbf{a} \cdot \mathbf{b} = 6x$
$\Rightarrow 2x^2 + 2\mathbf{a} \cdot \mathbf{b} = 6x$
$\Rightarrow \mathbf{a} \cdot \mathbf{b} = 3x - x^2$

Then, using $\mathbf{a} \cdot \mathbf{b} = |\mathbf{a}||\mathbf{b}|\cos\theta$,

$3x - x^2 = |\mathbf{a}||\mathbf{b}|\cos\theta$
$= x \times x \cos\theta$
$= x^2 \cos\theta$

Since $-1 \leq \cos\theta \leq 1$, $-x^2 \leq x^2 \cos\theta \leq x^2$ and so $-x^2 \leq 3x - x^2 \leq x^2$

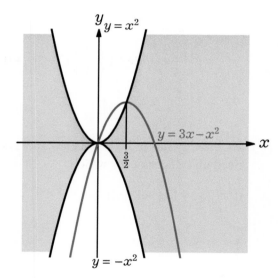

Figure 19S.27.1

From the graph on the GDC, $y = 3x - x^2$ is between $y = -x^2$ and $y = x^2$ for $x \geq \frac{3}{2} = 1.5$, so the smallest possible value of x is 1.5.

> **COMMENT**
>
> It is also possible to solve these inequalities without a calculator:
>
> $-x^2 \leq 3x - x^2 \leq x^2$
> $\Leftrightarrow 3x \geq 0$ and $2x^2 - 3x \geq 0$
> $\Leftrightarrow 3x \geq 0$ and $x(2x - 3) \geq 0$
> $\Leftrightarrow x \geq 0$ and $x \leq 0$ or $x \geq \frac{3}{2}$
> $\therefore x = 0$ or $x \geq \frac{3}{2}$
>
> so $x \geq \frac{3}{2}$ (as $x \neq 0$ is given)
>
> Nonetheless, it is still advisable to consult a graph when solving a quadratic inequality (such as $2x^2 - 3x \geq 0$); without a calculator, a sketch of the quadratic is needed.
>
> If it is unclear where the solutions to two inequalities both hold (such as $x \geq 0$ and $x \leq 0$ or $x \geq \frac{3}{2}$), highlight them on a number line and look for the region where they overlap:
>
>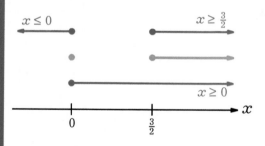
>
> **Figure 19S.27.2**

Long questions

1 a $P(\text{Daniel gets heads}) = \frac{1}{5}$

b $P(\text{Daniel gets tails, then Theo gets heads})$
$= \frac{4}{5} \times \frac{1}{5} = \frac{4}{25}$

c P(D gets tails, then T gets tails, then D gets heads)

$$= \frac{4}{5} \times \frac{4}{5} \times \frac{1}{5} = \frac{16}{125}$$

d The probabilities of Daniel winning on a particular throw are:

$$P(\text{first throw}) = \frac{1}{5}$$

$$P(\text{second throw}) = \left(\frac{4}{5}\right)^2 \left(\frac{1}{5}\right)$$

$$P(\text{third throw}) = \left(\frac{4}{5}\right)^4 \left(\frac{1}{5}\right)$$

$$\vdots$$

This is a geometric sequence with first term $\frac{1}{5}$ and common ratio $\left(\frac{4}{5}\right)^2$.

The probability of Daniel winning is S_∞, which is

$$S_\infty = \frac{\frac{1}{5}}{1-\left(\frac{4}{5}\right)^2}$$

$$= \frac{5}{25-16}$$

$$= \frac{5}{9}$$

e P(Theo wins) = 1 − P(Daniel wins)

$$= 1 - \frac{5}{9}$$

$$= \frac{4}{9}$$

> **COMMENT**
>
> Here we are assuming that the game eventually ends, so that there is no possibility of a draw. This is the case because the probability of no one winning after n throws is $P(n \text{ tails}) = \left(\frac{4}{5}\right)^n$, which tends to zero as $n \to \infty$.

f Let P(head) = p
Using the same argument as above,

$$P(\text{Daniel wins}) = \frac{p}{1-(1-p)^2}$$

If P(D wins) = 2 P(T wins), then since the two probabilities must add up to 1,

$$P(\text{D wins}) = \frac{2}{3}$$

$$\therefore \frac{p}{1-(1-p)^2} = \frac{2}{3}$$

$$3p = 2 - 2(1-2p+p^2)$$

$$2p^2 - p = 0$$

$$p(2p-1) = 0$$

So $p = \frac{1}{2}$ (as $p \neq 0$)

2 a $f(3) = 4$

$$\therefore f(f(3)) = f(4) = 6$$

b Translation by $\begin{pmatrix} 2 \\ 3 \end{pmatrix}$ results in the function $f(x-2)+3$, and reflection in the x-axis gives $g(x) = -f(x-2) - 3$

$$\therefore g'(x) = -f'(x-2)$$

and so $g'(2) = -f'(0) = -7$

3 a $f'(x) = 2 + \cos 2x - \sec^2 x$

b Stationary points occur where $f'(x) = 0$:

$$2 + \cos 2x - \sec^2 x = 0$$

$$2 + (2\cos^2 x - 1) - \left(\frac{1}{\cos x}\right)^2 = 0$$

$$2\cos^2 x + 2\cos^4 x - \cos^2 x - 1 = 0$$

$$2\cos^4 x + \cos^2 x - 1 = 0$$

c Solving the above equation for $\cos x$:

$$(2\cos^2 x - 1)(\cos^2 x + 1) = 0$$

$$\therefore 2\cos^2 x - 1 = 0 \quad (\text{as } \cos^2 x + 1 \neq 0)$$

$$\Rightarrow \cos x = \pm\frac{1}{\sqrt{2}}$$

$$\therefore x = \pm\frac{\pi}{4} \quad \text{as } x \in \left[-\frac{\pi}{2}, \frac{\pi}{2}\right]$$

i.e. there are two stationary points.

4 a $\sum_{r=0}^{\infty} a^r = 1 + a + a^2 + \ldots$

This is a geometric series with first term 1 and common ratio a.

$S_\infty = 1.5$

$\dfrac{1}{1-a} = 1.5$

$1.5 - 1.5a = 1$

$1.5a = 0.5$

$\therefore a = \dfrac{1}{3}$

b $1 - x + x^2 - x^3 + \ldots$ is a geometric series with first term 1 and common ratio $-x$.

Using the formula for S_∞,

$$1 - x + x^2 - x^3 + \ldots = \frac{1}{1-(-x)} = \frac{1}{1+x}$$

The series converges for $|-x| < 1$, i.e. $|x| < 1$

$\therefore k = 1$

c $\dfrac{d}{dx}\ln(1+x) = \dfrac{1}{1+x}$

$\therefore \ln(1+x) = \int \dfrac{1}{1+x} dx$

$= \int \left(1 - x + x^2 - x^3 + \ldots\right) dx$ from (b)

$= x - \dfrac{1}{2}x^2 + \dfrac{1}{3}x^3 - \dfrac{1}{4}x^4 + \ldots + c$

When $x = 0$,

$\ln(1) = c$

$\therefore c = 0$

Hence

$$\ln(1+x) = x - \frac{1}{2}x^2 + \frac{1}{3}x^3 - \frac{1}{4}x^4 + \ldots$$

COMMENT

This assumes that an infinite series can be integrated term by term, which is true in this case, although the formal proof requires techniques of mathematical analysis that are usually introduced at undergraduate level.

d Set $x = 0.1$ in (c):

$$\ln 1.1 \approx 0.1 - \frac{0.01}{2} + \frac{0.001}{3} - \frac{0.0001}{4}$$

$= (0.1 + 0.00033\ldots) - (0.005 + 0.000025)$

$= 0.10033\ldots - 0.005025$

$= 0.0953\ldots$

$= 0.095 \quad (3\text{DP})$

5 a For independent events,
$P(A \cap B) = P(A)P(B)$. But

$P(A) \times P(B) = 0.85 \times 0.60 = 0.51$

$P(A \cap B) = 0.55$

so A and B are not independent events.

b Require the probability that the building will not be completed on time (B') given that the materials arrive on time (A):

$P(B'|A) = 1 - P(B|A)$

$= 1 - \dfrac{P(A \cap B)}{P(A)}$

$= 1 - \dfrac{0.55}{0.85}$

$= 0.353 \quad (3\text{SF})$

c Let X be the number of hours worked by a random team member; then

$$X \sim N(42, \sigma^2)$$

First we need to find σ.

Let $Z = \dfrac{X-42}{\sigma} \sim N(0,1)$

$P(X > 48) = 10\%$

$\Rightarrow P\left(Z > \dfrac{48-42}{\sigma}\right) = 0.1$

$\Rightarrow \dfrac{6}{\sigma} = 1.2816$ (from GDC)

$\Rightarrow \sigma = 4.682$

Then

P(both plumbers work more than 40 hours)
$= P(X > 40) \times P(X > 40)$
$= 0.665^2$ (from GDC)
$= 0.443$

> **COMMENT**
>
> We assume that time is a continuous random variable, so 'more than 40 hours' means $X > 40$ rather than, say, $X \geq 41$ or $X > 40.5$.

6 a i Using long division, $\dfrac{2x+1}{x-3} = 2 + \dfrac{7}{x-3}$

As $x \to \infty$, $\dfrac{7}{x-3} \to 0$ and so $y = 2$ is a horizontal asymptote.

ii The vertical asymptote is where the denominator equals zero: $x - 3 = 0 \Rightarrow x = 3$

iii The two lines intersect at $(3, 2)$.

b When $x = 0$, $y = -\dfrac{1}{3}$

When $y = 0$, $2x + 1 = 0$, so $x = -\dfrac{1}{2}$

So the intersection points are $\left(0, -\dfrac{1}{3}\right)$ and $\left(-\dfrac{1}{2}, 0\right)$.

c This is a rational function with a shape like $\dfrac{1}{x}$:

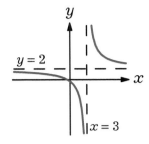

Figure 19L.6 Graph of $y = \dfrac{2x+1}{x-3}$

d Using the quotient rule,

$f'(x) = \dfrac{2(x-3) - 1(2x+1)}{(x-3)^2}$

$= \dfrac{-7}{(x-3)^2}$

When $x = 4$, the gradient of the tangent is $f'(4) = -7$ and the y-coordinate is $f(4) = 9$, so the equation of the tangent is

$y - 9 = -7(x - 4)$

$y = -7x + 37$

e Need to find the point T where $f'(x) = -7$:

$-\dfrac{7}{(x-3)^2} = -7$

$(x-3)^2 = 1$

$x - 3 = \pm 1$

$x = 4$ or 2

$x = 4$ is the point S, so T has $x = 2$. When $x = 2$, $y = -5$, so the coordinates of T are $(2, -5)$.

f The midpoint of $S(4, 9)$ and $T(2, -5)$ has coordinates

$x = \dfrac{4+2}{2} = 3, \quad y = \dfrac{9-5}{2} = 2$

which is the point $P(3, 2)$.

7 **a** $\dfrac{dy}{dx} = \dfrac{1}{x}$

When $x = e$, $\dfrac{dy}{dx} = \dfrac{1}{e}$

∴ the equation of the tangent at (e, 1) is

$y - 1 = \dfrac{1}{e}(x - e)$

$y = \dfrac{1}{e}x$

(0, 0) lies on this line because $0 = \dfrac{1}{e}(0)$

b Using the product rule on the first term:

$\dfrac{d}{dx}(x\ln x - x) = \left(1 \times \ln x + x \times \dfrac{1}{x}\right) - 1$

$= \ln x + 1 - 1$

$= \ln x$

c Shaded area

$= $ area of triangle with base e and height $1 - \int_1^e \ln x \, dx$

$= \dfrac{1}{2}(e)(1) - [x\ln x - x]_1^e$ (by part (b))

$= \dfrac{1}{2}e - \{(e\ln e - e) - (\ln 1 - 1)\}$

$= \dfrac{1}{2}e - e + e - 1$

$= \dfrac{1}{2}e - 1$

8 **a** The period of $f(x) = (\sin x)^2 \cos x$ is 2π, as this is the period of both sin and cos. (The period could be smaller, but it can be checked that $f(\pi) \neq f(0)$.)

> **COMMENT**
> You can also check the approximate period by drawing the graph on a GDC.

b Using the graph on a GDC, the minimum value of y is -0.385 and the maximum value is 0.385, so the range, to 1SF, is $-0.4 \leq y \leq 0.4$.

c **i** Using the product rule and chain rule:

$f'(x) = (2\sin x \cos x)\cos x + \sin^2 x(-\sin x)$

$= 2\sin x \cos^2 x - \sin^3 x$

ii At a maximum point (such as A),
$f'(x) = 0$

i.e. $2\sin x \cos^2 x - \sin^3 x = 0$

$\sin x (2\cos^2 x - \sin^2 x) = 0$

$\sin x = 0$ or $2\cos^2 x - \sin^2 x = 0$

$\sin x = 0$
$\Rightarrow x = 0, \pi$ or 2π

which cannot be the x-coordinate of A.

$2\cos^2 x - \sin^2 x = 0$
$\Rightarrow 2\cos^2 x - (1 - \cos^2 x) = 0$
$\Rightarrow 3\cos^2 x = 1$

Point A is the smallest positive solution of this equation, so

$\cos x = \sqrt{\dfrac{1}{3}}$

iii When $\cos x = \sqrt{\dfrac{1}{3}}$

$\sin^2 x = 1 - \cos^2 x$

$= 1 - \dfrac{1}{3}$

$= \dfrac{2}{3}$

So the maximum value is

$f(x) = \sin^2 x \cos x$

$= \dfrac{2}{3}\sqrt{\dfrac{1}{3}}$

$= \dfrac{2}{3\sqrt{3}}$

d At point B, $f(x) = 0$

$\therefore \sin^2 x \cos x = 0$

$\Rightarrow \sin x = 0$ or $\cos x = 0$

$\therefore x = 0, \dfrac{\pi}{2}, \pi, \ldots$

As B is the first positive x-intercept, it has $x = \dfrac{\pi}{2}$

e i Let $u = \sin x$; then

$\dfrac{du}{dx} = \cos x$

$dx = \dfrac{du}{\cos x}$

$\int \sin^2 x \cos x \, dx = \int u^2 \cos x \left(\dfrac{du}{\cos x} \right)$

$= \int u^2 \, du$

$= \dfrac{1}{3}u^3 + c$

$= \dfrac{1}{3}\sin^3 x + c$

ii The shaded area is between $x = 0$ and $x = \dfrac{\pi}{2}$

Area $= \int_0^{\frac{\pi}{2}} f(x) \, dx$

$= \left[\dfrac{1}{3}\sin^3 x \right]_0^{\frac{\pi}{2}}$

$= \dfrac{1}{3}$

f Point C is an inflection point, so it has $f''(x) = 0$:

$9(\cos x)^3 - 7\cos x = 0$

$\cos x (9\cos^2 x - 7) = 0$

$\Rightarrow \cos x = 0$ or $\pm\dfrac{\sqrt{7}}{3}$

$\cos x = 0$ at $x = \dfrac{\pi}{2}$, which is point B.

Point C has x-coordinate between 0 and $\dfrac{\pi}{2}$, so $\cos x > 0$

$\therefore x = \arccos\left(\dfrac{\sqrt{7}}{3}\right) = 0.491$ (3SF)

COMMENT

The equation can also be solved by drawing a graph or using the solver on the GDC.

9 a i p is the difference between the y-coordinates of the two graphs at the same value of x:

$$p = (10x+2) - (1+e^{2x})$$
$$= 10x+1-e^{2x}$$

ii At a maximum, $\frac{dp}{dx} = 0$:

$$10 - 2e^{2x} = 0$$
$$e^{2x} = 5$$
$$\therefore x = \frac{1}{2}\ln 5 = 0.805 \quad (3\text{SF})$$

b i $y = 1 + e^{2x}$
$$\Rightarrow e^{2x} = y - 1$$
$$\Rightarrow x = \frac{1}{2}\ln(y-1)$$
$$\therefore f^{-1}(x) = \frac{1}{2}\ln(x-1)$$

COMMENT

In the exam, there is no need to state the domain unless the question explicitly asks for it. In this case, the domain is $x > 1$.

ii $f(a) = 5$
$$\therefore a = f^{-1}(5)$$
$$= \frac{1}{2}\ln 4$$
$$= \ln\sqrt{4}$$
$$= \ln 2$$

c $V = \pi \int_0^{\ln 2} (1+e^{2x})^2 \, dx$

10 a $y = 3x - 5$
$$\Rightarrow y + 5 = 3x$$
$$\Rightarrow x = \frac{y+5}{3}$$
$$\therefore f^{-1}(x) = \frac{x+5}{3}$$

b $g^{-1}(f(x)) = (3x-5)+2 = 3x-3$

c $(f^{-1} \circ g)(x) = (g^{-1} \circ f)(x)$
$$\frac{x+3}{3} = 3x - 3$$
$$x + 3 = 9x - 9$$
$$8x = 12$$
$$x = \frac{3}{2}$$

d i $h(x) = \frac{3x-5}{x-2}$

The graph is similar in shape to $y = \frac{1}{x}$, with:

- vertical asymptote where $x - 2 = 0$, i.e. $x = 2$
- horizontal asymptote at $y = \frac{3}{1} = 3$
- x-intercept where $3x - 5 = 0$, i.e. $x = \frac{5}{3}$
- y-intercept at $y = \frac{-5}{-2} = \frac{5}{2}$

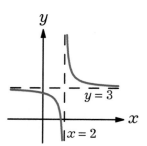

Figure 19L.10.1 Graph of $y = \frac{3x-5}{x-2}$

COMMENT

This graph could just be drawn on the GDC; the vertical asymptote will be immediately apparent, and the horizontal asymptote can be found by looking at large values of x on the GDC. However, it is important to be able to use the above process to sketch a function's graph, in case you need to do so on the non-calculator paper.

ii The asymptotes are $x = 2$ and $y = 3$.

e i $\int h(x)\,dx = \int 3 + \dfrac{1}{x-2}\,dx$

$= 3x + \ln|x-2| + c$

ii $\int_3^5 h(x)\,dx = \left[3x + \ln|x-2|\right]_3^5$

$= (15 + \ln 3) - (9 + \ln 1)$

$= 6 + \ln 3$

f

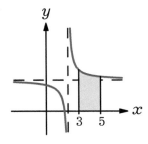

Figure 19L.10.2 The area represented by $\int_3^5 h(x)\,dx$

11 a Using the product rule and the chain rule:

Let $u = (x-a)^p$, $v = (x-b)^q$; then

$\dfrac{du}{dx} = p(x-a)^{p-1}$, $\dfrac{dv}{dx} = q(x-b)^{q-1}$

$\therefore \dfrac{dy}{dx} = p(x-a)^{p-1}(x-b)^q + q(x-b)^{q-1}(x-a)^p$

Stationary point where $\dfrac{dy}{dx} = 0$:

$p(x-a)^{p-1}(x-b)^q + q(x-b)^{q-1}(x-a)^p = 0$

$(x-a)^{p-1}(x-b)^{q-1}\left[p(x-b) + q(x-a)\right] = 0$

$(x-a)^{p-1}(x-b)^{q-1}\left[(p+q)x - (pb+qa)\right] = 0$

Since $a < x < b$, $x-a$ and $x-b$ are not zero.

$\therefore \dfrac{dy}{dx} = 0$ only when $(p+q)x - (pb+qa) = 0$

i.e. $x = \dfrac{pb+qa}{p+q}$

b $y = (x-a)^2(x-b)^3$ and $a < b$

This is a polynomial of degree 5 with zeros a and b. The factor $(x-a)$ is squared so the curve touches the x-axis there. The factor $(x-b)$ is cubed, so the curve crosses the x-axis at b. (In fact, it looks like x^3 near $x = b$, so this is a point of inflection.) According to part (a), the graph has one stationary point.

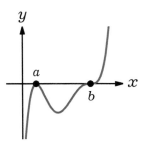

Figure 19L.11.1 Graph of $y = (x-a)^2(x-b)^3$

c The shape of the graph depends on the parity of p and q, because this determines both the behaviour for large negative x and whether the graph crosses or touches the x-axis.

We know that:
- $y = 0$ for $x = a$ and $x = b$ only
- when x is large and positive, y is large and positive

Also:
- when x is large and negative, y is positive if $p+q$ is even and negative if $p+q$ is odd
- the graph touches the x-axis at $x = a$ if p is even, and crosses it if p is odd
- the graph touches the x-axis at $x = b$ if q is even, and crosses it if q is odd

The possible shapes of the graph are:

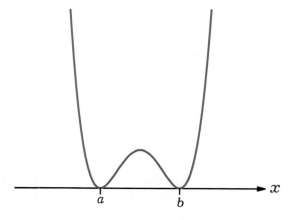

Figure 19L.11.2 Graph of $y = (x-a)^p (x-b)^q$ for p even, q even, $p+q$ even

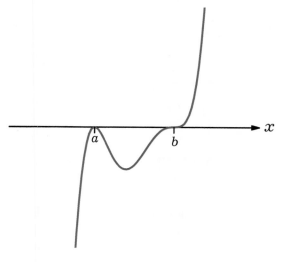

Figure 19L.11.3 Graph of $y = (x-a)^p (x-b)^q$ for p even, q odd, $p+q$ odd

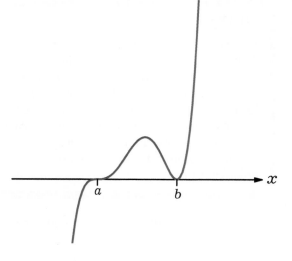

Figure 19L.11.4 Graph of $y = (x-a)^p (x-b)^q$ for p odd, q even, $p+q$ odd

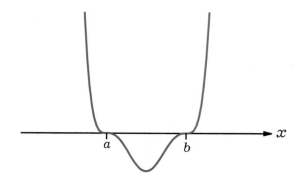

Figure 19L.11.5 Graph of $y = (x-a)^p (x-b)^q$ for p odd, q odd, $p+q$ even

From these graphs, it can be seen that the stationary point is a local maximum in the first and third cases, i.e. when q is even.